Springer Series on Touch and Haptic Systems

More information about this series at http://www.springer.com/series/8786

Claudio Pacchierotti

Cutaneous Haptic Feedback in Robotic Teleoperation

Claudio Pacchierotti
Department of Advanced Robotics
Istituto Italiano di Tecnologia (IIT)
Genoa
Italy

Additional material to this book can be downloaded from http://extras.springer.com.

ISSN 2192-2977 ISSN 2192-2985 (electronic)
Springer Series on Touch and Haptic Systems
ISBN 978-3-319-25455-5 ISBN 978-3-319-25457-9 (eBook)
DOI 10.1007/978-3-319-25457-9

Library of Congress Control Number: 2015954583

Springer Cham Heidelberg New York Dordrecht London

Printed on acid-free paper

Springer International Publishing AG Switzerland is part of Springer Science+Business Media
(www.springer.com)

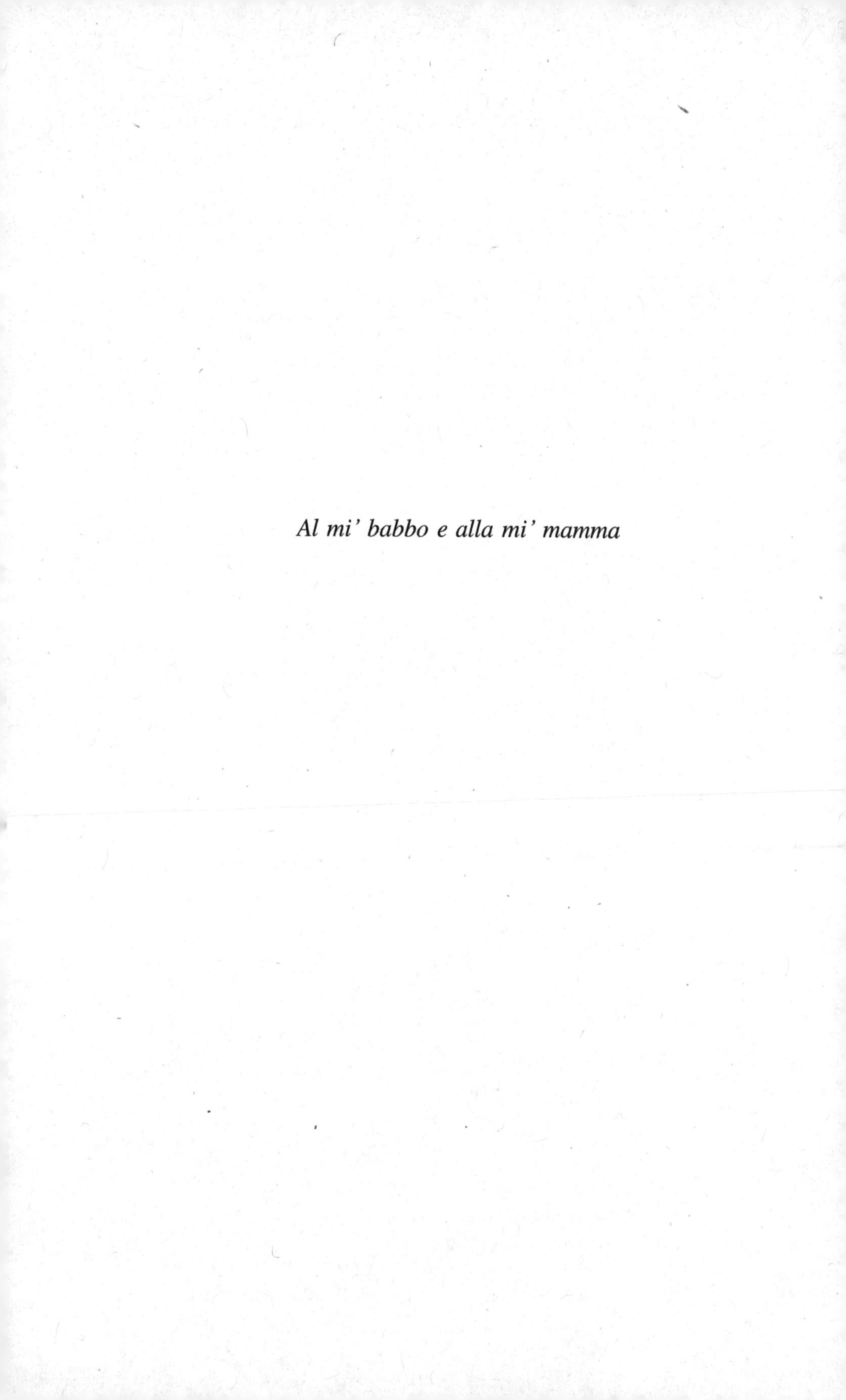

Al mi' babbo e alla mi' mamma

Series Editors' Foreword

This is the 12th volume of the "Springer Series on Touch and Haptic Systems," which is published in collaboration between **Springer** and the **EuroHaptics Society**.

Cutaneous Haptic Feedback in Robotic Teleoperation is focused on analyzing haptic feedback by considering tactile and kinesthetic components. A new approach named "sensory subtraction," which represents a transformation of the interaction forces before being reflected on a user, is described and validated in telemanipulation scenarios, such as telesurgery and industrial robotics. Feedback of cutaneous interaction is a strong challenge in telerobotics since it is related to soft object deformations that can easily provoke unstable bilateral systems. Examples of how to solve these problems are provided by common telesurgery tasks such as needle teleoperation. In this book, engineering and perception issues are considered in order to properly display interaction forces during telemanipulation tasks on a user. This book is organized into two parts and six chapters. The first part is focused on cutaneous cues feedback and the second part is concerned with the integration of cutaneous and kinesthetic cues.

Dr. Claudio Pacchierotti has received the EuroHaptics 2014 Ph.D. award. In recognition of this award, he was invited to publish his work in the Springer Series on Touch and Haptic Systems. Pacchierotti's thesis was selected in an international competition from a large number of excellent theses all focused on the sense of touch that were defended in 2014. This volume of the "Springer Series on Touch and Haptic Systems" provides a state of the art on how to properly reflect telemanipulation forces.

September 2015

Manuel Ferre
Marc O. Ernst
Alan Wing

Foreword

Cutaneous deformation feedback is generated by haptic devices designed to deform the skin through some kind of mobile platform, and they are intended to provide no kinesthetic stimuli. Such type of feedback can simulate very well the shape and softness of virtual and remote objects. The information conveyed by this type of cutaneous feedback is very rich and, at the same time, it does not affect the stability of teleoperation systems. Moreover, the deformation provided is very close to the one experienced during direct interaction with real objects. In fact, the cutaneous deformation provided approximates the initial deformation of the skin during the interaction with an object, leaving out the subsequent kinestethic part. This is the intrinsic nature of the cutaneous feedback considered in the book of Claudio Pacchierotti.

The application of cutaneous feedback to robotic teleoperation is a natural consequence of the richness of information this type of feedback conveys and of the intrinsic stability it guarantees thanks to the low energy involved in the stimuli and the design of our cutaneous devices. The application of cutaneous feedback to teleoperation was so natural that Claudio and I chose to refer to this approach not as another sensory substitution technique but as a novel sensory subtraction method. In fact, haptic feedback can be considered to be provided to the human operator through a combination of cutaneous and kinaesthetic stimuli. For this reason, providing cutaneous stimuli only can be also seen as "subtracting" kinaesthetic feedback from the full haptic interaction. With the term "sensory subtraction" we want therefore to highlight the fact that we are removing the part of the interaction that can affect the stability and safety of our systems, i.e., kinesthetic feedback, leaving only the cutaneous cues. However, we had a lot of concerns about the use of the term "subtraction", which might give the impression of somehow degrading the feedback information. Nonetheless, in the end, the algebraic interpretation of the term subtraction—cutaneous is equal to haptic minus kinesthetic—won. We hope that the reader will understand our choice and appreciate this naming as much as we do.

The first part of the book elegantly introduces the sensory subtraction idea and its application in a 2 needle insertion experiment, a peg-in-a-hole task, and a robot-assisted surgical scenario. From this first part we also understand another interesting feature of cutaneous haptic feedback: devices providing only cutaneous stimuli can be significantly smaller and less bulky than their kinesthetic counterparts, enabling us to easily integrate them in complex existing systems, such as the da Vinci Surgical System, where the sensory subtraction idea takes the shape of a thimble directly attached on the master console. The second part of the book explains how the synergistic exploitation of both cutaneous and kinesthetic feedbacks improves the performance of teleoperation systems with force reflection while guaranteeing their stability and safety.

Dr. Pacchierotti's admirable book establishes the foundations of cutaneous haptic feedback in robotic teleoperation and wonderfully addresses its applications in both surgical and industrial robotics. The book provides a comprehensive experimental evaluation of cutaneous haptic feedback in various teleoperation scenarios, proving that cutaneous feedback can truly be a viable and effective approach to force feedback in robotic teleoperation.

September 2015 Prof. Domenico Prattichizzo

Acknowledgments

First and foremost I would like to express my sincere gratitude to my Ph.D. advisor, Prof. Domenico Prattichizzo, for the continuous support of my study and research, for his motivation and contagious enthusiasm through all these years. His guidance helped me to grow both as a research scientist and a man, from the very first project together (when I was still a freshman!) to my Ph.D. thesis.

I wish to thank my fellow labmates for helping me during these years of hard work and for being such good friends, inside and outside the lab. I wish them all the best as they finish their own degrees and head out into the world.

I am extremely grateful to Prof. Katherine J. Kuchenbecker, for her hospitality, help, and support during my visit to her lab. She has contributed invaluably to my research and personal growth. I could have not invested better my months abroad, both from a personal and professional point of view. Together with her, I wish to thank my labmates at Penn, who made me feel immediately part of their group. My experience in Philadelphia has been made great also by the extraordinary people I met at the International House, who significantly contributed to my happiness and positive state of mind.

I am grateful to Prof. Giulio Rosati and Prof. Sarthak Misra for their hospitality and help during my visits in Padova and Enschede, respectively. The interdisciplinary nature of our projects broaden my view on the topics of haptic, robotics, and rehabilitation.

I would also like to thank all the developers around the world who gave their time to provide me and the community with incredible free software tools, such as GCC, ROS, GNOME, Linux, Inkscape, and OpenShot. Without these tools my work would have never been possible.

My parents, Gabriella and Giampiero, receive my deepest gratitude and love for their dedication and the many years of support during my studies.

Finally, I wish to thank all my good friends in Siena, who have never stopped cheering me up since high school times.

The research leading to these results has received funding from the European Union Seventh Framework Programme FP7/2007–2013 under grant agreement n°601165 of the project “WEARHAP—WEARable HAPtics for humans and robots”.

Contents

About the Author

Claudio Pacchierotti received the B.S., M.S., and Ph.D. degrees from the University of Siena, Italy in 2009, 2011, and 2014, respectively. He was an exchange student at the Karlstad University, Sweden in 2010. He spent the first seven months of 2014 visiting the Penn Haptics Group at the University of Pennsylvania, Philadelphia, USA, which is part of the General Robotics, Automation, Sensing, and Perception (GRASP) Laboratory. He also visited the Department of Innovation in Mechanics and Management of the University of Padua and the Institute for Biomedical Technology and Technical Medicine (MIRA) of the University of Twente in 2013 and 2014, respectively. He received the 2014 EuroHaptics Best Ph.D. Thesis Award for the best doctoral thesis in the field of haptics. He is currently a postdoctoral researcher at the Department of Advanced Robotics of the Italian Institute of Technology, Genova, Italy. His research deals with robotics and haptics, focusing on cutaneous force feedback techniques, wearable devices, and haptics for robotic surgery.

Introduction

ἐν μὲν γὰρ ταῖς ἄλλαις λείπεται πολλῶν τῶν ξῴων,
κατὰ δὲ τὴν ἁφὴν πολλῷ τῶν ἄλλων διαφερόντως
ἀκριβοῖ. διὸ καὶ φρονιμώτατόν ἐστι τῶν ξῴων.
σημεῖον δὲ τὸ καὶ ἐν τῷ γένει τῶν ἀνθρώπων παρὰ τὸ
αἰσθητήριον τοῦτο εἶναι εὐφυεῖς και ἀφυεῖς, παρ' ἄλλο
δὲ μηδέν· οἱ μὲν γὰρ σκληρόσαρκοι ἀφυεῖς τὴν
διάνοιαν, οι δὲ μαλακόσαρκοι εὐφυεῖς.

Ἀριστοτέλης, Περὶ Ψυχῆς

In the other senses man is inferior to many of the animals, but in the sense of touch he is far superior to the rest. And to this he owes his superior intelligence. This may be seen from the fact that it is this organ of sense and nothing else which makes all the difference in the human race between the natural endowment of man and man. Hard-skinned men are naturally deprived of intelligence, but the soft-skinned ones are naturally gifted with it.

Aristotle, *De Anima*

Aristotle in his *De anima* defines the sense of touch as the most precise sense of the human race, the one that makes humans the most intelligent animals among all the others living on Earth. He also asserts that humans with a more developed sense of touch are the ones naturally gifted with the most intelligence. In 1960, Frank Geldard advocated that the sense of touch was a "neglected sense of communication," observing that, while visual and auditory sensing systems were respectively superior at spatial and temporal discrimination, the somatosensory system was capable of both [1, 2]. More recently, science journalist Natalie Angier wrote in the New York Times that "Biologically, chronologically, allegorically and delusionally, touch is the mother of all sensory systems. It is an ancient sense in evolution: even the simplest single-celled organisms can feel when something brushes up against them and will respond by nudging closer or pulling away [3]."

However, surprisingly, while evolution saw force senses in the limbs and skin of primitive animals long before the exteroceptors of light and sound developed, human technologies have developed in a reverse fashion [4]. First, in the second half of the nineteenth century, Thomas Edison claimed the invention of the phonograph. It recorded sound onto a tinfoil sheet phonograph cylinder and could both record and reproduce sounds. Then it was the time of moving images, reproduced along with sounds, which foreran modern television systems. Along with the rapid and widespread of audio and vision technologies, we also witnessed a drastic change in their form factor and target usage, from big and heavy machines to small and lightweight objects. Think, for instance, at the first hi-fi speakers, and compare them to the latest portable music players. The same applies to technologies made to reproduce video signals. From the first cathode ray tube televisions, nowadays companies are developing flexible organic light-emitting diode screens, which can be easily shipped in rolls.

What about the sense of touch?

What about devices able to reproduce the feeling of touching remote objects, similarly to how speakers and display screens reproduce remote sounds and videos?

Unlike speakers and display screens, artificial devices reproducing the feeling of touching remote objects have started to spread only recently. The science that took up this challenge is called *haptics*. As K.J. Kuchebecker pointed out [5], the adjective "haptic" is just a more formal synonym for the term "touch based," which origins from the Greek word ἁπτικός, meaning "able to touch or grasp." Etymology aside, A.M. Okamura links the word *haptics* to something more familiar to the layman and tells The Washington Post that "haptics is to touch as optics is to sight" [6]. The same expression stands also out in the home page of W. Provancher's haptics group at Utah [7], where I saw it for the first time. After years of vainly trying to explain to my family and friends what haptics is, this is now my opening line.

Haptics technology thus refers to our ability of designing artificial systems able to sense and transmit the several pieces of information we get from feeling the real world, similarly to how video cameras and display screens register information and feed them to our eyes, respectively. Haptics technology enabling humans to touch remote objects has typically been used in robotics for teleoperation. Telerobotic systems usually involve a slave robot, which interacts with the remote environment, and a master console, operated by a human operator. The slave robot reproduces the movements of the operator, who in turn needs to observe the remote environment with which the robot is interacting. The latter can be achieved by a combination of visual and haptic cues that flow from the environment to the operator. Visual feedback is already available in several popular telerobotic systems (e.g., the Intuitive Surgical da Vinci Si and the Space Shuttle Canadarm), but current teleoperated systems have very limited haptic feedback. This omission is related to many different factors. One of the most relevant is the negative effect that haptic feedback has on the stability of teleoperation systems. Haptic force feedback can in fact lead to undesired oscillations of the system, which interfere with the operation and may be dangerous for both the environment and the human operator [8, 9]. In this respect, cutaneous feedback has recently received great attention; delivering ungrounded sensory cues to the operator's skin conveys rich information and does not affect the stability of these teleoperation systems [8, 10, 11].

This book presents my contribution to the field of haptics and robotics, collecting all the work I have done toward my Ph.D. degree at the University of Siena and at the Italian Institute of Technology (January 2012–December 2014). It addresses the challenge of providing effective cutaneous feedback in robotic teleoperation, with the objective of achieving the highest degree of transparency while guaranteeing the stability, and thus the safety, of the considered systems. The book is divided in two main parts: cutaneous-only approaches (Part I) and mixed cutaneous–kinesthetic approaches (Part II).

Part I presents teleoperation systems that provide only cutaneous cues to the operator, thus guaranteeing the highest degree of safety. As mentioned before, in fact, cutaneous feedback does not affect the stability of teleoperation systems. We called this approach *sensory subtraction*, in contrast to sensory substitution, as it subtracts the kinesthetic part of the full haptic interaction to leave only the cutaneous cues. The sensory subtraction approach is best suitable for those scenarios where the safety of the system is paramount, e.g., robotic surgery. In this respect, Chap. 2 presents an application of the sensory subtraction idea in a 1 degree of freedom (DoF) simulated needle insertion task. Chap. 3 presents an application of the sensory subtraction idea in a more challenging remote peg-in-hole task, both in simulated and real environments. Finally, Chap. 4 presents an application of the sensory subtraction idea in a remote palpation task using the da Vinci Surgical System.

On the other hand, Part II presents teleoperation systems that provide mixed cutaneous and kinesthetic cues to the operator. In this respect, Chap. 5 presents a teleoperation system with haptic feedback wherein cutaneous cues are used to compensate for the temporary reduction of haptic feedback necessary to satisfy certain stability conditions. This mixed approach aims at improving the transparency of cutaneous-only systems described in Part I. Finally, Chap. 6 presents a teleoperation system where mixed kinesthetic and vibrotactile navigation feedback helps the operator in the steering of a bevel-tipped flexible needle in a tissue phantom.

References

1. F.A. Geldard, Some neglected possibilities of communication. Science **131**, 1583–1588 (1960)
2. J. Pasquero, Survey on Communication Through Touch. Tech. Rep. TR-CIM, vol. 6 (Center for Intelligent Machines-McGill University, 2006)
3. N. Angier, Primal, acute and easily duped: our sense of touch, New York Times, p. D2, 2008
4. T.B. Sheridan, Further musings on the psychophysics of presence, in Proc. IEEE International Conference on Systems, Man, and Cybernetics, vol. 2 (1994), pp. 1073–1077

5. K.J. Kuchenbecker, Characterizing and controlling the high-frequency dynamics of haptic interfaces, PhD thesis, Stanford University, 2006
6. J. Garreau, Touch screen technology, from iphones to ovens, continues to push forward, The Washington Post, 2008. [Online]. http://www.washingtonpost.com/wp-dyn/content/article/2008/12/14/AR2008121402455.html
7. W. Provancher. Haptics and Embedded Mechatronics Laboratory (University of Utah, Salt Lake City, Utah, 2014)
8. D. Prattichizzo, C. Pacchierotti, G. Rosati, Cutaneous force feedback as a sensory subtraction technique in haptics. IEEE Trans. Haptics. 5(4), 289–300 (2012)
9. N. Diolaiti, G. Niemeyer, F. Barbagli, J.K. Salisbury, Stability of haptic rendering: discretization, quantization, time delay, and coulomb effects. IEEE Trans. Rob. 22(2), 256–268 (2006)
10. L. Meli, C. Pacchierotti, D. Prattichizzo, Sensory subtraction in robot-assisted surgery: fingertip skin deformation feedback to ensure safety and improve transparency in bimanual haptic interaction. IEEE Trans. Biomed. Eng. 61(4) 1318–1327 (2014)
11. W. McMahan, J. Gewirtz, D. Standish, P. Martin, J.A. Kunkel, M. Lilavois, A. Wedmid, D.I. Lee, K.J. Kuchenbecker, Tool contact acceleration feedback for telerobotic surgery. IEEE Trans. Haptics. 4(3) 210–220 (2011)

Part I
Force Feedback via Cutaneous Cues Only

Chapter 1
Sensory Subtraction in Teleoperation: Substituting Haptic Force with Cutaneous Stimuli

Abstract Cutaneous haptic feedback is considered an elegant solution to enhance the performance of robotic teleoperation systems while guaranteeing their safety. Delivering ungrounded cutaneous cues to the human operator conveys in fact rich information and does not affect the stability of the control loop. For this reason, this first part of the book presents a novel feedback approach to robotic teleoperation, that substitutes full haptic feedback with cutaneous stimuli only, with the objective of achieving high levels of performance while guaranteeing the stability of the considered systems. This technique is named *sensory subtraction*, in contrast to sensory substitution, as it subtracts the destabilizing kinesthetic part from the full haptic interaction to leave only cutaneous cues. Similarly to sensory substitution, this kind of feedback is expected to make the teleoperation system intrinsically stable, since the cutaneous force applied does not affect the position of the master device, thus opening the feedback loop. This chapter introduces the concept of sensory subtraction, and it briefly reviews the literature on ungrounded cutaneous devices and how they have been employed in robotic teleoperation.

1.1 Introduction

A teleoperator is a machine that allows humans to sense and mechanically manipulate objects at a distance, by virtually relocating the operators at a place other than their true location [1]. It includes artificial sensors to perceive the remote environment, actuators to be able to move into it, and network channels to communicate with the human operator. In order to perform complex tasks, teleoperators may also include artificial devices (e.g., arms or hands) to apply forces and perform mechanical work on the environment.

If the human operator receives sufficient information about the slave system and the remote environment, he will feel as if he is actually present at the remote site: this condition is commonly referred to as telepresence [2, 3]. Achieving a good illusion of telepresence is a matter of technology. If the teleoperator transmits sufficient information to the users, displayed in a sufficiently articulated way, the illusion of telepresence can be compelling [1, 4]. The primary tool to achieve this objective is

C. Pacchierotti, *Cutaneous Haptic Feedback in Robotic Teleoperation*,
Springer Series on Touch and Haptic Systems, DOI 10.1007/978-3-319-25457-9_1

providing a *transparent* implementation of the teleoperation system. Transparency can be in turn defined as the correspondence between the master and slave positions and forces [5], or as the match between the impedance of the environment and the one perceived by the operator [6]. Achieving telepresence thus hinges upon transmitting different types of information from the remote environment to the human operator. These pieces of information usually consist of a combination of visual and haptic stimuli. Visual feedback is already widely employed in commercial robotic teleoperation systems (e.g., the da Vinci Si Surgical System, Intuitive Surgical, USA) while haptics is still missing from the market. One of the reasons for this omission is the negative effect that haptic feedback may have on the stability of teleoperation systems. Indeed, stability of teleoperation systems with force reflection can be significantly affected by communication latency in the loop, hard contacts, relaxed grasps, and many other destabilizing factors that dramatically reduce the effectiveness of haptics in teleoperation [5] (see Fig. 1.1a). Despite stability issues, haptic stimuli still play a fundamental role in enhancing the performance of teleoperation systems in terms of completion time of a given task [7–9], accuracy [8], peak [10, 11] and mean force [9, 11]. In medicine, haptic feedback has been shown to improve performance in fine microneedle positioning [12], telerobotic catheter insertion [13], suturing simulation [8], cardiothoracic procedures [14], and cell injection systems [15]. For this reason, guaranteeing the stability of teleoperation systems with haptic feedback has always been a great challenge. It can be addressed by either avoiding the use of actuators on the master console or by designing appropriate control systems. In the former case, no force feedback is provided through the master end-effector, making the teleoperation system intrinsically stable and thus guaranteeing the highest degree of safety [9, 16, 17]. In the absence of any actuator on the master console, an effective approach to still provide information about the forces exerted at the slave side is *sensory substitution*: it consists of replacing haptic feedback with other forms of feedback, such as vibrotactile [18], auditory, and/or visual feedback [19]. Kitagawa et al. [19] evaluated the effect of substituting haptic feedback with visual and auditory cues during a teleoperated surgical knot-tying task. Forces applied while using these sensory substitution modalities more closely approximated suture tensions achieved under ideal haptic conditions (i.e., hand ties) than forces applied without such feedback. Schoonmaker and Cao [18] introduced vibrotactile stimulation on the foot to substitute haptic feedback during minimally invasive surgery (MIS) procedures. Results show improved ability in differentiating tissue softness. However, although sensory substitution techniques guarantee the stability of these systems, the stimuli provided may be very different from the ones being substituted (e.g., a beep sound instead of force feedback). Therefore, sensory substitution often shows worse performance than that achieved employing unaltered force feedback [16, 17].

In this respect, cutaneous feedback has recently received great attention in the haptic and robotic research fields; delivering ungrounded cutaneous cues to the operator's skin conveys rich information and does not affect the stability of the teleoperation system [16, 17, 20]. Cutaneous stimuli are sensed by mechanoreceptors in the skin and they are useful to recognize the local properties of objects such as shape, edges,

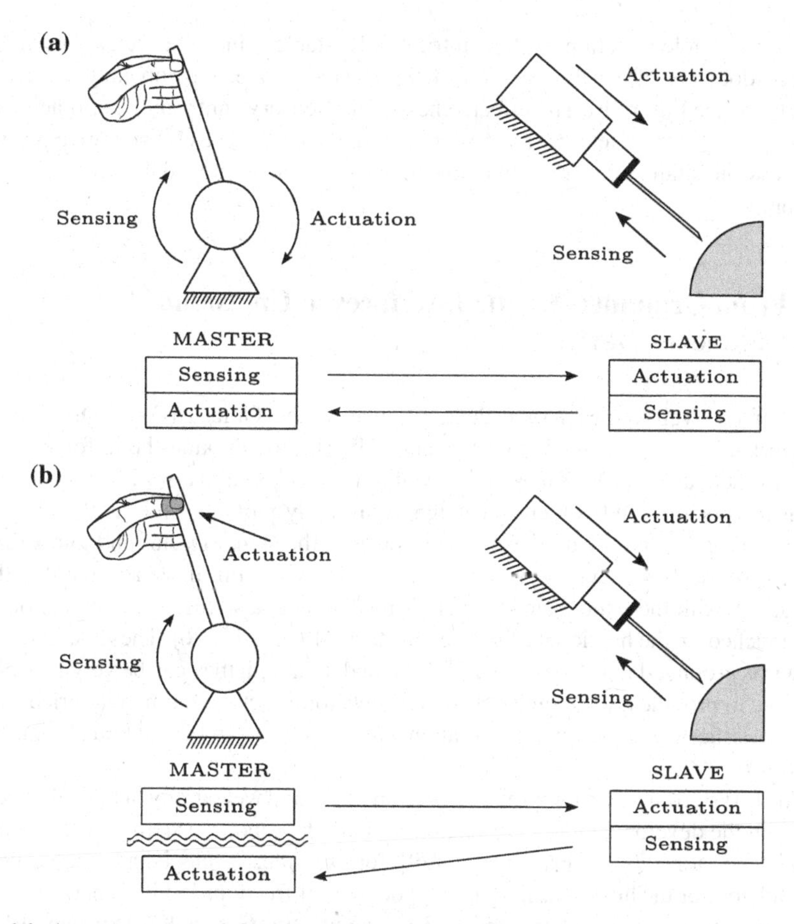

Fig. 1.1 A common approach for teleoperation versus our sensory subtraction technique. The loss of realism due to providing cutaneous force only is often a price worth paying to gain a great improvement in the safety of the system. **a** A common approach for teleoperation systems. The force fed back to the user is directly applied on the end-effector of the master device, which is also in charge of steering the slave robot. A control action is needed to avoid instability. **b** Teleoperation system employing cutaneous feedback *only*. Force feedback is applied to the fingertips of the operator and the loop is intrinsically stable

embossings and recessed features. This is possible, principally, thanks to a direct measure of intensity and direction of contact forces and to the encoding of the force spatial distribution over the fingertip [21, 22].

Similarly to sensory substitution, we present here a novel feedback approach that substitutes haptic feedback with cutaneous feedback in teleoperation. We named this technique *sensory subtraction*, in contrast to sensory substitution, as it subtracts the kinesthetic part from the full haptic interaction to leave only the cutaneous cues (see also Sect. 1.2). Similarly to sensory substitution, we expect this kind of feedback to make the teleoperation system intrinsically stable, since the cutaneous force applied does not affect the position of the master device, thus opening the feedback loop (see Fig. 1.1b). The effectiveness of the sensory subtraction approach will be tested in different teleoperation scenarios in Chaps. 2–4. I will now give a brief overview on cutaneous devices and how they have been employed in robotic teleoperation.

1.2 From Grounded Haptic Interfaces to Ungrounded Cutaneous Devices

Most of the well-known haptic devices, such as the Omega (Force Dimension, Switzerland) or the Phantom (3D Systems, USA), provide kinesthetic force feedback to their users [23]. Kinesthetic feedback provides humans with information about the position and velocity of neighboring body parts, as well as the applied force and torque, mainly by means of receptors in the skin, muscles and joints [23–25]. However, these devices provide also *cutaneous* stimuli, if we assume that the interaction with the remote environment is mediated by a stylus, a ball, or any other tool attached on the haptic interface's end-effector [9, 16, 17, 26]. These devices are known as grounded interfaces (see Fig. 1.2a) and, although they can be very accurate and able to provide a wide range of forces, their form factor, commercial price, and stability issues when used in teleoperation, challenged researchers to find alternative solutions.

The first step, aiming at improving the form factor and wearability of these devices, has been the development of exoskeletons, where the robotic system is worn by the human operator [27–29]. Frisoli et al. [29], for example, developed a tendon driven exoskeleton for the human arm. It has five degrees of freedom and it is characterized by a workspace similar to the one of the human arm. CyberGlove Systems LLC developed a hand exoskeleton able to provide force feedback to all five fingers of the hand simultaneously. It can provide up to 12 N per finger and has a mass of 0.5 kg. Exoskeletons are able to provide both kinesthetic and cutaneous feedback, similarly to grounded haptic devices. However, their kinesthetic feedback is limited to the body-grounded base. For example, in the hand exoskeleton sketched in Fig. 1.2b the user receives kinesthetic feedback on the finger but not on the arm, in contrast to what is happening in Fig. 1.2a with a grounded haptic system. Another drawback of body-grounded haptics is that two forces are applied to the user: the contact force simulating the interaction and an *undesired* reaction force, which counterbalances the first one (blue and red arrows in Fig. 1.2b, respectively). In grounded haptics this force is counterbalanced by the ground and thus not felt by the user. A good design principle to address this issue is to distribute this reaction force onto a large contact surface, thus making it less perceivable than the one simulating the contact interaction [28, 30].

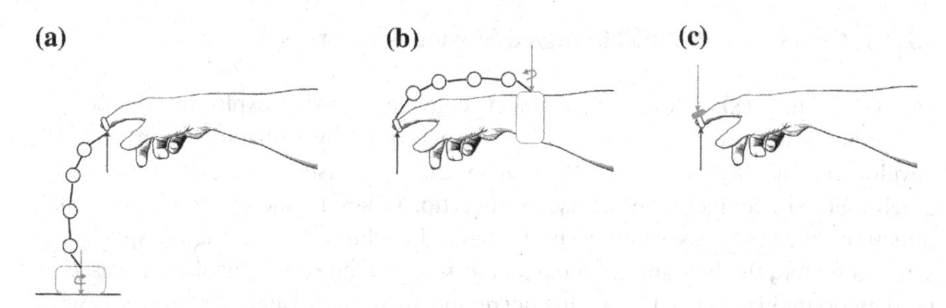

Fig. 1.2 From grounded to ungrounded haptics. Grounded haptic devices (**a**), exoskeletons (**b**) and ungrounded interfaces (**c**). The contact force simulating the interaction and the *undesired* reaction force, which counterbalances the first one, are shown in *blue* and *red*, respectively. Ungrounded devices will be used in the next chapters to provide cutaneous feedback while guaranteeing the stability of the teleoperation system (color figure online)

As the body-grounded base moves toward the point of application of the interaction force, the kinesthetic feedback the system is able to provide reduces. At the extreme end of this process we find ungrounded haptic systems, where the body-grounded base is placed next to the point of application of the force simulating the interaction. An example is the device sketched in Fig. 1.2c, where the base is located on the nail and the interaction force is applied at the fingertip. Ungrounded haptic devices can be considered as providing only cutaneous feedback [26, 30].

However, reducing the force feedback provided to solely cutaneous stimuli should not be seen as a problem, but as an opportunity. As mentioned before, in fact, cutaneous devices can be used in teleoperation to provide force feedback in a safe way. Moreover, the simplified form factor of these devices can dramatically reduce their market cost, opening the doors to a broader diffusion of haptic technologies. Finally, the high wearability of these cutaneous devices enables the operator to use them *together* with commercial grounded haptic interfaces, with the objective of providing decoupled cutaneous and kinesthetic stimuli. This latter approach will be discussed in Chap. 5.

1.2.1 Ungrounded Cutaneous Devices

Several types of cutaneous haptic devices have been developed in the past years. A comprehensive review on the topic is out of the scope of this book, but I will give here a brief overview on those devices using tactile technologies similar to the ones employed in the next chapters. A more comprehensive overview on cutaneous technologies used in teleoperation scenarios is presented in Sect. 1.3.

1.2.1.1 Cutaneous Stimuli Through a Moving Tractor

One of the first examples of ungrounded cutaneous device exploiting the design guidelines discussed in Sect. 1.2 is the one presented by Minamizawa et al. [31], developed to display the weight of virtual objects. It consists of two motors that move a belt that is in contact with the user's fingertip. When the motors spin in opposite directions, the belt presses into the user's fingertip, while when the motors spin in the same direction, the belt applies a tangential force to the skin. This device was also used by Prattichizzo et al. [32] to display remote tactile experiences: an instrumented glove worn by a human sensed interaction forces at the remote environment, and the above cutaneous device presented those sensations to the user. Gleeson et al. [33] introduced a two-degree-of-freedom (2-DoF) cutaneous device that laterally stretches the skin of the fingertip using a 7 mm hemispherical tractor. Its two RC servo motors and compliant flexure stage can move the tractor along any path in the plane of the fingerpad. The device is capable of rendering 1 mm of displacement at arbitrary orientations within a plane, with a rate of 5 mm/s. The device has been also used to guide a human user navigating an unknown space [34]. Wang and Quek [35] developed a system to remotely transmit tactile sensations; their two armbands each contain tactile pressure sensors and six shape memory alloy (SMA) actuators. Each device's sensors register the pattern and amplitude of applied force, and the other armband attempts to generate the same haptic sensations via the SMA actuators. More recently, Prattichizzo et al. [30] presented a wearable 3-DoF cutaneous device for interaction with virtual and remote environments. It consists of two platforms: one is located on the back of the finger, supporting three small DC motors, and the other is in contact with the volar surface of the fingertip. The motors shorten and lengthen three cables to move the platform toward the user's fingertip and re-angle it to simulate contacts with arbitrarily oriented surfaces. The direction and amount of the force reflected to the user is changed by properly controlling the cable lengths. The actuators used for the device prototype are three 0615S Faulhaber motors, with planetary gear-heads having 16:1 reduction ratio. The maximum stall torque of the motor is 3.52 m Nm. Three force-sensing resistors near the platform vertices measure the fingertip contact force for closed-loop control. Pacchierotti et al. [26] presented an improved version of the same device that achieves higher accuracy by using motors with encoders and a single force sensor. It consists again of two platforms connected by three wires. Three small electrical motors, equipped with position encoders, control the length of the wires, moving the mobile platform toward the fingertip. One force sensor is placed at the platform's center, in contact with the finger pulp. A modified version of this device will be employed in Chaps. 2 and 5.

Although these cutaneous devices have been successfully employed in various scenarios, their end-effectors *always* contact the finger pad. They thus cannot provide the sensation of breaking and making contact with virtual and remote surfaces, cues that are known to be important to tactile interaction [25, 36]. Provancher et al. [37] designed a novel contact location display to overcome this limitation; it includes a roller that translates along as well as makes and breaks contact with the user's fingertip. Kuchenbecker et al. [38] employed a similar principle to create a non-actuated fingertip device that provides the user with the cutaneous sensation of making and breaking contact with virtual surfaces. When it is attached to a traditional haptic interface, force feedback deflects small internal springs and brings a shell into contact with the user's fingertip. Frisoli et al. [39] achieved a similar effect by creating a finger-mounted thimble that moves a 5-DoF flat contact plate around the fingertip. The device can be also attached to the end-effector of a grounded haptic interface to combining the characteristics of an encounter-type haptic system with display of contact surface orientation. Solazzi et al. [40] developed a 3-DoF wearable cutaneous display to render virtual slanted surfaces. Four motors are placed on the forearm and two cables for each actuated finger transmit the motor torque to the fingertips. More recently, Pacchierotti et al. [9] presented a 3-DoF cutaneous device for remote tactile interaction. Its design is similar to the ones described in [26, 30], but it adds three springs to enable the platform to make and break contact with the fingertip. A modified version of this device will be employed in Chaps. 3 and 4.

Although these devices greatly improved the rendering of remote and virtual environments by displaying cutaneous sensations, they have all been employed in scenarios that treat the cutaneous interaction as a point force rather than spatially distributed sensations. The device presented by Kuchenbecker et al. [38], for example, is attached to a single-point grounded haptic device; Prattichizzo et al. [32] measured remote forces through only one force sensor; and the authors of [9, 26, 30] evaluated the virtual environment contact forces as though they were applied at one contact point. On the one hand, this simplified approach makes these haptic systems very easy to control, with only a few input parameters and simple force sensing systems. On the other hand, this approach cannot correctly represent the wide range of sensations the fingertip may encounter during a real interaction, since it does fully account for the spatial distribution of cutaneous receptors [24, 25]. Moreover, most of the aforementioned devices make use of skin deformation models to determine the actuator inputs needed to apply a given force. For example, Pacchierotti et al. [9] used an isotropic elastic model of the fingertip, assuming a linear relationship between platform displacement and resultant wrench, while Gleeson et al. [33] characterized the interaction between their device and the skin through an experiment with seven human subjects. Although they have served well, such models do not guarantee accurate delivery of the desired force on the user's fingertip. This problem will be discussed and addressed in Chap. 4, where the algorithm presented accounts for the spatial distribution of cutaneous receptors and directly map sensed deformations to motor inputs, without using any skin deformation model.

1.2.1.2 Vibrotactile Cutaneous Stimuli

In addition to the above mentioned type of cutaneous devices, there is also a growing interest in vibrotactile cutaneous feedback. Vibrations have been in fact successfully employed to provide navigation information and contact acceleration feedback in many scenarios. Erp et al. [41], for example, explored the possibility of presenting navigation information through a vibrating waist belt. Results indicate the usefulness of vibrotactile cues for navigation purposes as well as for situational awareness in multi-tasks environments. A similar device has been developed in [42], where a haptic belt was integrated with a directional sensor and a Global Positioning System (GPS), and employed as an intuitive navigation system. Traylor and Tan [43] presented a vibrating wearable device able to provide directional information on the user's back. The tactile display consists of a single tractor strapped to the volar side of the user's forearm. An accelerometer is placed on top of the tractor to record its displacement during signal delivery. Lieberman and Breazeal [44] presented a 5-DoF arm suit able to guide the motion of the wearer by providing solely vibrotactile feedback. The suit is composed of eight vibrotactile actuators distributed throughout the right arm, whose frequency and amplitude are independently controlled. Kim et al. [45] presented a vibrotactile display to provide safety information to drivers. The device is placed on top of the foot and it is composed of a 5×5 array of vibrating motors. Huisman et al. [46] presented a device to remotely transmit vibrations to the forearm. It consists of a soft pressure-sensitive input device and of a vibrating interface. A touch to one's own forearm is felt as a vibration on the forearm of another person. Culbertson et al. [47] presented a new method for creating haptic texture models from data recorded during natural and unconstrained motions. An accelerometer mounted on a stylus records the vibrations occurring as it is dragged across a real surface. A force sensor in the stylus measures the applied normal force. The computer then generates a synthetic texture vibration in real time, and a voice-coil actuator mounted on the stylus transmits these vibrations to the user's hand. A vibrating haptic bracelet has been used by Scheggi et al. [48] for human-robot interaction in leader-follower formation tasks. The bracelet consists of three vibrating motors providing the user with information about the robot formation. Vibrotactile feedback has been also employed in robot-assisted surgery: Schoonmaker and Cao [18] demonstrated that vibrotactile stimulation is a viable substitute for force feedback in minimally invasive surgery, enhancing surgeons' ability to control the forces applied to tissue and differentiate its softness in a simulated tissue probing task. More recently, McMahan et al. [20] developed a sensing and actuating device for the da Vinci Surgical System able to provide vibrotactile feedback of tool contact accelerations (see Sect. 1.3). Eleven surgeons tested the system and expressed a significant preference for the inclusion of vibrotactile feedback.

Vibrotactile cutaneous stimuli are employed in Chap. 4, where a vibrotactile motor provides the operator with vibrations as recorded by a biomimetic tactile sensor at the slave side, and in Chap. 6, where mixed kinesthetic-vibrotactile stimuli provide clinicians with navigation feedback during the steering of flexible needles in soft tissue.

1.3 Cutaneous Feedback in Robotic Teleoperation

The ungrounded cutaneous devices presented in the previous section have been successfully employed in many different scenarios, taking advantage of their high level of wearability, inexpensiveness, and safety. In this section I will give an overview on those cutaneous technologies when used in robotic teleoperation, focusing on medical applications. The intrinsic stability guaranteed by cutaneous feedback is in fact opening new and promising scenarios in the medical field, where the safety of the system is a paramount and non-negotiable requirement. Cutaneous feedback is also appealing because it can be implemented without significant modifications of the master interface or the telerobotic system's control algorithms.

It is currently impossible to create a cutaneous display that can perfectly match the wide array of sensations a human fingertip can feel, so researchers have tried to design devices able to emulate sensations for specific applications. I will thus categorize the considered systems based on the type of cutaneous stimuli provided and the type of experimental evaluation carried out. Biggs and Srinivasan [27] provide an in-depth review of haptic interfaces and define a list of four primitives of tactile sensations: normal indentation, lateral skin stretch, relative tangential motion, and vibration. The large variety of tactile sensations humans experience when interacting with virtual or remote objects can be considered combinations of these few building blocks [27]. To put my work in perspective, this overview will also include the systems presented in this book.

1.3.1 Normal Indentation

Normal indentation displays convey cutaneous stimuli through one or multiple moving tractors, providing spatially distributed tactile information through the indentation of the tractors into the skin. Peeters et al. [49] claimed that this is "the preferred stimulation method for tactile feedback systems to be used in minimally invasive surgery".

An example of cutaneous devices providing this kind of cutaneous stimuli are pin-array displays. Already in 1995, Howe et al. [50] developed a pin-array display to rectify the deficit of cutaneous feedback in the operating theater. The display raises pins against the human fingertip skin to approximate a desired shape. It is composed of a 6×4 array of pins actuated via shape-memory alloy (SMA) wires, with a center-to-center pin spacing of 2.1 mm. The authors validated the system by carrying out an experiment of remote palpation. Two pin-array cutaneous devices were attached to the end-effector of two grounded kinesthetic interfaces, so to track the position of the fingertips and provide additional kinesthetic feedback. At the remote site, two tactile sensors, attached to the end-effector of two other grounded kinesthetic interfaces, registered the forces exerted on the environment. The motion of the remote interfaces was directly controlled by the human operator from the master site. Subjects were asked to locate a hard rubber cylinder inside a block of foam rubber. The task simulated the key aspects of medical procedures where tumors are localized with palpation. Using cutaneous feedback, subjects were able to locate the tumor with an error of 1 mm or less in over 50 % of the trials, and with an error of 3 mm or less 95 % of the time. When cutaneous feedback was not available, the mean absolute error was over 13 mm. Similarly, Feller et al. [51] built a 6×6 pin array for remote palpation actuated via RC servo motors. The center-to-center pin spacing is 3 mm and the diameter of each pin is 1 mm. In the experimental evaluation, subjects controlled with the right hand an instrumented remote manipulator using a Phantom interface and felt the tactile display with the left index finger. The task consisted in locating hard lumps in a remote soft tissue phantom by exploring it with the telemanipulator. The Phantom was in charge of providing kinesthetic feedback and the pin-array display was in charge of providing cutaneous stimuli. Cutaneous feedback was found again to be an effective solution to provide force feedback for remote palpation. Also Kim et al. [52] developed a pin-array tactile display for tumor palpation. It provides spatially distributed tactile information through a 6×8 pin array actuated via piezoelectric motors. The center-to-center pin spacing is 1.7 mm and the diameter of each pin is 0.7 mm. The cutaneous display is attached to the end-effector of a Phantom haptic interface to track the position of the fingertip and provide additional kinesthetic feedback. The experimental evaluation was carried out in a deformable 3-D virtual environment simulating soft human tissue. Twenty subjects were asked to detect the position of four tumors hidden in the simulated tissue. Results show that providing cutaneous and kinesthetic feedback significantly increased the performance of the palpation task with respect to providing solely kinesthetic feedback through the Phantom interface. More recently, Ottermo et al. [53] presented a remote palpation instrument for laparoscopic surgery. It consists of a 4×8 pin array actuated via DC motors. The center-to-center pin spacing is 2.7 mm and the diameter of each pin is 2 mm. Subjects controlled a laparoscopic grasper with the right hand, and felt the tactile display with their left index finger, similarly to [51]. A cardboard box was used as a laparoscopic simulator. On top of the box five holes were cut. These functioned as trocars and were covered with rubber to make them more flexible. Five objects were hidden in latex finger cots and placed in the cardboard box. Subject was asked to rank the hidden objects from the smallest to the biggest using the laparoscopic system with and without cutaneous feedback. Cutaneous feedback improved the performance of all but one subject.

Similarly to pin arrays, another popular set of cutaneous systems providing stimuli via normal indentations of mobile tractors are pneumatic balloon-based systems. King et al. [54, 55] developed a modular pneumatic tactile feedback system to improve surgical performance of the da Vinci Surgical System. The system is composed of a piezoresistive force sensor mounted onto the Cadiere graspers of the robot and two pneumatic balloon-based tactile displays mounted on the robot's master console. Sixteen novices and four experts were asked to perform several repetitions of a peg transfer task, with and without the cutaneous feedback system. Results showed that the presence of cutaneous feedback significantly reduced grip force during robotic manipulation and that the reduction of force was not sustainable once the feedback was removed. More recently, Li et al. [56] extended this approach to three fingers, presenting a compact pneumatic system for robot-assisted minimally invasive surgery. It simulates soft tissue stiffness by changing the pressure of three air balloons placed on the index, middle and ring fingers. The authors carried out a tumor palpation experiment in a virtual environment. The task consisted in recognizing the stiffness of three tumors hidden in soft human tissue relying only on the cutaneous stimuli provided by the pneumatic system. All participants were able to feel the presence of the simulated tumors.

Pin and air balloon arrays provide spatially distributed tactile information through multiple moving tractors. This means that, in addition to normal stresses, they can also provide tactile information by changing the contact area between the skin and the display. Bicchi et al. [57] investigated the possibility of substituting cutaneous stimuli for softness discrimination with information on the rate of spread of the contact area between the finger and the display as the contact force increases. They presented a device for implementing such cutaneous feedback and a practical application to a mini-invasive surgery tool. The cutaneous display consists of ten cylinders of different radii in telescopic arrangement. Regulated air pressure acts on one end of the cylinders, and the operator finger probes the other end of the display. The length of the cylinders is arranged so that, when no forces are applied by the operator, the active surface of the display is a stepwise approximation of a cone. As the force applied increases, the contact area with the skin increases. The experimental evaluation consisted in measuring the capability of subjects to recognize five different remote items by their stiffness. The authors compared recognition rates using direct exploration, a grounded kinesthetic interface, and the proposed cutaneous display. Average recognition rates using the cutaneous display (75 %) outperformed kinesthetic feedback (37 %) and provides results comparable with direct exploration of the remote items (87 %). Similarly, Gwilliam et al. [58] described an adjustable aperture air-jet pneumatic lump display that directs a thin stream of pressurized air through an aperture onto the finger pad. Increasing the air pressure increases the normal force provided at the fingertip, while increasing the air-jet aperture increases the contact area. The display is designed to produce the sensation of a lump with minimal hardware requirements.

The cutaneous devices employed in Chaps. 2–4 apply cutaneous stimuli through a 3-DoF moving platform. Although the platform applies a combination of normal and shear forces to the skin, the normal component is always present. For this reason I included these devices is this section. Sect. 3.2.2 discusses in details the relationship between the normal and shear forces provided by this type of devices.

Chapter 2 presents a teleoperation system where a customized version of the device presented in [30] is used together with a grounded haptic interface to steer a needle in soft tissue. The cutaneous device provides planar fingertip deformation feedback to the operator's fingertip through a 3-DoF mobile platform controlled by three DC motors. The cutaneous device substitutes the full haptic feedback of the grounded device with cutaneous stimuli only, according to the sensory subtraction paradigm. Experiments in a virtual environment show that the proposed cutaneous-only approach, other than being intrinsically stable, improves the teleoperation performance with respect to sensory substitution of haptic feedback with visual feedback [16]. A similar system has been presented by Meli et al. [17]. The authors used four prototypes of the abovementioned 3-DoF fingertip device together with two 7-DoF grounded haptic interfaces. The cutaneous devices again substitute the full haptic feedback of the grounded devices with cutaneous stimuli only. The authors tested the performance of the system in a virtual manipulation task similar to the Peg Board experiment of the da Vinci Skills Simulator. Results showed cutaneous stimuli to be more effective than no force feedback at all. Moreover, cutaneous feedback outperformed two popular sensory substitution technique: visual and auditory feedback in substitution of haptic feedback.

The ungrounded device employed in these two works is quite effective and, thanks to its form factor, it can be easily used together with popular grounded haptic interfaces. This feature will be exploited in Chap. 5 to combine the kinesthetic feedback provided by a grounded interface with the cutaneous feedback provided by our cutaneous devices. However, when it is not necessary to preserve the ability of using other interfaces, such a compact form factor is not required. Chapter 3 presents a teleoperation system where two prototypes of the device presented in [9] are used in a paradigmatic remote peg-in-hole task. Each cutaneous device provides planar fingertip deformation feedback to the operator's fingertip through a 3-DoF mobile platform controlled by three servo motors. Experiments in virtual and real environments show that the proposed cutaneous-only approach improves the teleoperation performance with respect to providing no force feedback at all [59]. An augmented version of the same device is used in Chap. 4, where we present a cutaneous feedback system for the da Vinci surgical robot. Designed to provide planar fingertip deformation *and* vibration cues to the surgeon, the system is composed of a BioTac tactile sensor mounted to one of the robot's slave tools and a cutaneous display device attached to the corresponding master controller. The cutaneous device provides planar fingertip deformation feedback through a 3-DoF mobile platform controlled by three servo motors, as in Chap. 3, and vibrotactile feedback through a vibrotactile motor attached to the same platform. We tested the system in a soft tissue palpation task using the da Vinci Surgical System. Results show that providing cutaneous feedback significantly improved the task performance with respect to not providing force feedback at all. However, providing planar fingertip deformation through the mobile platform was sufficient to outperform the condition not providing any force feedback. Adding vibrotactile feedback did not significantly improve the registered performance [60–62].

Although the results presented in Chap. 4 are very promising, the proposed tactile system enables the operator to palpate the tissue only with one finger. However, rather than poking, pinching better matches clinical practice [63–65]. For this reason, we are currently investigating technological solutions to provide cutaneous stimuli to more than one finger on the da Vinci master console.

1.3.2 Lateral Skin Stretch and Relative Tangential Motion

Lateral skin stretch is a feedback modality in which a shear force is applied to the skin. It is an interesting type of cutaneous feedback because it exploits the high sensitivity of human skin to tangential stretches and can provide the user with directional information. Schorr et al. [66] evaluated the potential of skin stretch feedback for robotic teleoperation systems. They presented a fingertip skin stretch feedback device that imposes tangential skin stretch proportionally to the intended level of force feedback. The authors carried out an experiment to determine the ability of subjects in discriminating between virtual surfaces of different stiffness using skin stretch feedback. The cutaneous device was attached to the end-effector of a Phantom haptic interface to track the position of the fingertip and provide additional kinesthetic feedback. Results show that users' stiffness discrimination capability using solely skin stretch was comparable to that of using kinesthetic feedback. Furthermore, larger skin stretch cues were perceived as portraying greater stiffness without any advance training.

Skin stretch and tangential motion cutaneous stimuli can be combined together to provide the illusion of slippage. Murphy et al. [67] developed a 2-DoF haptic display capable of reproducing the sensation of sliding contact at the fingertip. It is composed of a spherical ball supported under the user's fingertip. The user contacts a portion of the ball through an aperture in the mechanism housing. Two orthogonal wheels actuate the ball to create relative motion between the surface of the ball and the fingertip. The device is then connected to a Phantom haptic interface through a passive gimbal attachment, so to provide additional kinesthetic stimuli. More recently, Westebring–van der Putten et al. [68] presented a cylindrical rotating device able to provide slip sensations to the fingertip. The height and diameter of the cylinder is 25 mm. In order to understand the influence of skin stretch and tangential motion feedback on laparoscopic grasp control, the authors carried out two experiments. In the first experiment, four groups learned a single-handed laparoscopic lifting task. Three groups received feedback (visual, haptic, or a combination of them) on slip and excessive pinch force. In the second experiment, the authors considered a two-handed task. While the right hand was lifting the object as in the first experiment, a button had to be pushed with a tool held with the left hand. Subjects received either cutaneous or no feedback on grasp forces. In the second experiment, subjects who received cutaneous feedback could control their pinch force significantly better than subjects who did not receive force feedback.

1.3.3 Vibration

In addition to the above mentioned types of cutaneous feedback, there is also a growing interest in vibrotactile cutaneous feedback. As discussed in Sect. 1.2.1.2, vibrations have been employed in many different scenarios to provide navigation information and contact acceleration feedback. In Chap. 4 we present a cutaneous feedback system for the da Vinci surgical robot designed to provide planar fingertip deformation *and* vibration cues to the surgeon. Although adding vibrations did not improve the performance of the considered palpation task, most of the subjects preferred the condition providing vibrotactile feedback with respect to the one providing cutaneous feedback only through the moving platform. Vibrotactile feedback was also used in the teleoperation hand system developed by Kontarinis et al. [69].

Master and slave manipulators are identical two-fingered hands with 2 DoF in each finger. The system uses a direct-drive, parallel linkage design, which minimizes friction, backlash, and moving mass. Sensors in the fingertips of the remote manipulator measure the vibrations generated during task execution, and high frequency vibrotactile displays relay these vibrations to the human operator. Providing vibrotactile feedback significantly improved the results of the considered teleoperation experiments.

The most notable work employing vibrotactile feedback in robot-assisted surgery has been presented by McMahan et al. [20]. They have developed a sensing and actuating device for the da Vinci Surgical System able to provide auditory and vibrotactile feedback of tool contact accelerations. The authors carried out an experimental evaluation with eleven surgeons. They were asked to use the augmented da Vinci robot to perform three in vitro manipulation tasks under four different feedback conditions: with no acceleration feedback, with audio feedback, with vibrotactile feedback, and with both audio and vibrotactile feedback. Subjects preferred the conditions providing vibrotactile cutaneous feedback of tool contact accelerations.

References

1. T.B. Sheridan, Teleoperation, telerobotics and telepresence: a progress report. Control Eng. Pract. **3**(2), 205–214 (1995)
2. T.B. Sheridan, Telerobotics. Automatica **25**(4), 487–507 (1989)
3. T.B. Sheridan, *Telerobotics, Automation, and Human Supervisory Control* (The MIT press, Cambridge, 1992)
4. J.V. Draper, D.B. Kaber, J.M. Usher, Telepresence. Hum. Factors: J. Hum. Factors Ergon. Soc. **40**(3), 354–375 (1998)
5. K. Hashtrudi-Zaad, S.E. Salcudean, Transparency in time-delayed systems and the effect of local force feedback for transparent teleoperation. IEEE Trans. Robot. Autom. **18**(1), 108–114 (2002)
6. D.A. Lawrence, Stability and transparency in bilateral teleoperation. IEEE Trans. Robot. Autom. **9**(5), 624–637 (1993)
7. M.J. Massimino, T.B. Sheridan, Teleoperator performance with varying force and visual feedback. Hum. Factors: J. Hum. Factors Ergon. Soc. **36**(1), 145–157 (1994)
8. L. Moody, C. Baber, T.N. Arvanitis, Objective surgical performance evaluation based on haptic feedback. Stud. Health Technol. Inform. **85**, 304–310 (2002)
9. C. Pacchierotti, F. Chinello, M. Malvezzi, L. Meli, D. Prattichizzo, Two finger grasping simulation with cutaneous and kinesthetic force feedback. Haptics: Percept. Devices Mobil. Commun. **7282**, 373–382 (2012)
10. B. Hannaford, Task-level testing of the JPL-OMV smart end effector, in *Proceedings of the Workshop on Space Telerobotics*, vol. 2 (1987)
11. C.R. Wagner, N. Stylopoulos, R.D. Howe, The role of force feedback in surgery: analysis of blunt dissection, in *Proceedings of the Symposium of Haptic Interfaces for Virtual Environment and Teleoperator Systems* (2002), pp. 68–74
12. S.E. Salcudean, S. Ku, G. Bell, Performance measurement in scaled teleoperation for microsurgery, in *Proceedings of the First Joint Conference on Computer Vision, Virtual Reality and Robotics in Medicine and Medial Robotics and Computer-Assisted Surgery* (1997), pp. 789–798

13. A. Kazi, Operator performance in surgical telemanipulation. Presence: Teleoperators Virtual Environ. **10**(5), 495–510 (2001)
14. C.W. Kennedy, T. Hu, J.P. Desai, A.S. Wechsler, J.Y. Kresh, A novel approach to robotic cardiac surgery using haptics and vision. Cardiovasc. Eng. **2**(1), 15–22 (2002)
15. A. Pillarisetti, M. Pekarev, A.D. Brooks, J.P. Desai, Evaluating the effect of force feedback in cell injection. IEEE Trans. Autom. Sci. Eng. **4**(3), 322–331 (2007)
16. D. Prattichizzo, C. Pacchierotti, G. Rosati, Cutaneous force feedback as a sensory subtraction technique in haptics. IEEE Trans. Haptics **5**(4), 289–300 (2012)
17. L. Meli, C. Pacchierotti, D. Prattichizzo, Sensory subtraction in robot-assisted surgery: fingertip skin deformation feedback to ensure safety and improve transparency in bimanual haptic interaction. IEEE Trans. Biomed. Eng. **61**(4), 1318–1327 (2014)
18. R.E. Schoonmaker, C.G.L. Cao, Vibrotactile force feedback system for minimally invasive surgical procedures, in *Proceedings of the IEEE International Conference on Systems, Man and Cybernetics*, vol. 3 (2006), pp. 2464–2469
19. M. Kitagawa, D. Dokko, A.M. Okamura, D.D. Yuh, Effect of sensory substitution on suture-manipulation forces for robotic surgical systems. J. Thorac. Cardiovas. Surg. **129**(1), 151–158 (2005)
20. W. McMahan, J. Gewirtz, D. Standish, P. Martin, J.A. Kunkel, M. Lilavois, A. Wedmid, D.I. Lee, K.J. Kuchenbecker, Tool contact acceleration feedback for telerobotic surgery. IEEE Trans. Haptics **4**(3), 210–220 (2011)
21. I. Birznieks, P. Jenmalm, A.W. Goodwin, R.S. Johansson, Encoding of direction of fingertip forces by human tactile afferents. J. Neurosci. **21**(20), 8222–8237 (2001)
22. K.O. Johnson, The roles and functions of cutaneous mechanoreceptors. Curr. Opin. Neurobiol. **11**(4), 455–461 (2001)
23. V. Hayward, O.R. Astley, M. Cruz-Hernandez, D. Grant, G. Robles-De-La-Torre, Haptic interfaces and devices. Sens. Rev. **24**(1), 16–29 (2004)
24. B.B. Edin, N. Johansson, Skin strain patterns provide kinaesthetic information to the human central nervous system. J. Physiol. **487**(1), 243–251 (1995)
25. B.B. Edin, L. Ascari, L. Beccai, S. Roccella, J.-J. Cabibihan, M.C. Carrozza, Bio-inspired sensorization of a biomechatronic robot hand for the grasp-and-lift task. Brain Res. Bull. **75**(6), 785–795 (2008)
26. C. Pacchierotti, A. Tirmizi, D. Prattichizzo, Improving transparency in teleoperation by means of cutaneous tactile force feedback. ACM Trans. Appl. Percept. **11**(1), 4:1–4:16 (2014)
27. S.J. Biggs and M. Srinivasan, Haptic interfaces, in *Handbook of Virtual Environments*, (2002), pp. 93–116
28. M. Bergamasco, B. Allotta, L. Bosio, L. Ferretti, G. Parrini, G.M. Prisco, F. Salsedo, G. Sartini, An arm exoskeleton system for teleoperation and virtual environments applications, in *Proceedings of the IEEE International Conference on Robotics and Automation* (1994), pp. 1449–1454
29. A. Frisoli, F. Rocchi, S. Marcheschi, A. Dettori, F. Salsedo, M. Bergamasco, A new force-feedback arm exoskeleton for haptic interaction in virtual environments, in *Proceedings of the World Haptics* (2005), pp. 195–201
30. D. Prattichizzo, F. Chinello, C. Pacchierotti, M. Malvezzi, Towards wearability in fingertip haptics: a 3-dof wearable device for cutaneous force feedback. IEEE Trans. Haptics **6**(4), 506–516 (2013)
31. K. Minamizawa, S. Fukamachi, H. Kajimoto, N. Kawakami, S. Tachi, Gravity grabber: wearable haptic display to present virtual mass sensation, in *Proceedings of the ACM Special Interest Group on Computer Graphics and Interactive Techniques Conference* (2007), 8-es
32. D. Prattichizzo, F. Chinello, C. Pacchierotti, K. Minamizawa, Remotouch: a system for remote touch experience, in *Proceedings of the IEEE International Symposium on Robots and Human Interactive Communications* (2010), pp. 676–679
33. B.T. Gleeson, S.K. Horschel, W.R. Provancher, Design of a fingertip-mounted tactile display with tangential skin displacement feedback. IEEE Trans. Haptics **3**(4), 297–301 (2010)

34. R.L. Koslover, B.T. Gleeson, J.T. de Bever, W.R. Provancher, Mobile navigation using haptic, audio, and visual direction cues with a handheld test platform. IEEE Trans. Haptics **5**(1), 33–38 (2012)
35. R. Wang, F. Quek, Touch & talk: contextualizing remote touch for affective interaction, in *Proceedings of the International conference on Tangible, Embedded, and Embodied Interaction* (2010), pp. 13–20
36. G. Westling, R.S. Johansson, Responses in glabrous skin mechanoreceptors during precision grip in humans. Exp. Brain Res. **66**(1), 128–140 (1987)
37. W.R. Provancher, M.R. Cutkosky, K.J. Kuchenbecker, G. Niemeyer, Contact location display for haptic perception of curvature and object motion. Int. J. Robot. Res. **24**(9), 691–702 (2005)
38. K.J. Kuchenbecker, D. Ferguson, M. Kutzer, M. Moses, A.M. Okamura, The touch thimble: providing fingertip contact feedback during point-force haptic interaction, in *Procedings of the Symposium on Haptic Interfaces for Virtual Environment and Teleoperator Systems* (2008), pp. 239–246
39. A. Frisoli, M. Solazzi, F. Salsedo, M. Bergamasco, A fingertip haptic display for improving curvature discrimination. Presence: Teleoperators Virtual Environ. **17**(6), 550–561 (2008)
40. M. Solazzi, A. Frisoli, M. Bergamasco, Design of a cutaneous fingertip display for improving haptic exploration of virtual objects, in *Proceedings of the IEEE International Symposium on Robots and Human Interactive Communications* (2010), pp. 1–6
41. J.B.F. van Erp, H.A.H.C.V. Veen, C. Jansen, T. Dobbins, Waypoint navigation with a vibrotactile waist belt. ACM Trans. Appl. Percept. **2**(2), 106–117 (2005)
42. K. Tsukada, M. Yasumura, Activebelt: belt-type wearable tactile display for directional navigation, in UbiComp. Ubiquitous Comput. **2004**, 384–399 (2004)
43. R. Traylor, H.Z. Tan, Development of a wearable haptic display for situation awareness in altered-gravity environment: some initial findings, in *Proceedings of the Symposium on Haptic Interfaces for Virtual Environment and Teleoperator Systems* (2002), pp. 159–164
44. J. Lieberman, C. Breazeal, Tikl: development of a wearable vibrotactile feedback suit for improved human motor learning. IEEE Trans. Robot. **23**(5), 919–926 (2007)
45. H. Kim, C. Seo, J. Lee, J. Ryu, S. Yu, S. Lee, Vibrotactile display for driving safety information, in *Proceedings of the IEEE Intelligent Transportation Systems Conference* (2006), pp. 573–577
46. G. Huisman, A.D. Frederiks, B. Van Dijk, B. Kröse, D. Heylen, Self touch to touch others: designing the tactile sleeve for social touch, in *Proceedings of the ACM International Conference on Tangible, Embedded and Embodied Interaction* (2013)
47. H. Culbertson, J.M. Romano, P. Castillo, M. Mintz, K.J. Kuchenbecker, Refined methods for creating realistic haptic virtual textures from tool-mediated contact acceleration data, in *Proceedings of the IEEE Haptics Symposium* (2012), pp. 385–391
48. S. Scheggi, F. Chinello, D. Prattichizzo, Vibrotactile haptic feedback for human-robot interaction in leader-follower tasks, in *Proceedings of the International Conference on Pervasive Technologies Related to Assistive Environments* (2012), p. 51
49. K. Peeters, M. Sette, P. Goethals, J. Vander Sloten, H. Van Brussel, Design considerations for lateral skin stretch and perpendicular indentation displays to be used in minimally invasive surgery, in Haptics: Perception, Devices and Scenarios (2008), pp. 325–330
50. R.D. Howe, W.J. Peine, D.A. Kantarinis, J.S. Son, Remote palpation technology. IEEE Eng. Med. Biol. Mag. **14**(3), 318–323 (1995)
51. R.L. Feller, C.K. L. Lau, C.R. Wagner, D.P. Perrin, R.D. Howe, The effect of force feedback on remote palpation, in *Proceedings of the IEEE International Conference on Robotics and Automation* (2004), pp. 782–788
52. S.-Y. Kim, K.-U. Kyung, J. Park, D.-S. Kwon, Real-time area-based haptic rendering and the augmented tactile display device for a palpation simulator. Adv. Robot. **21**(9), 961–981 (2007)
53. M.V. Ottermo, Ø. Stavdahl, T.A. Johansen, A remote palpation instrument for laparoscopic surgery: design and performance. Minim. Invasive Ther. Allied Technol. **18**(5), 259–272 (2009)
54. C.-H. King, M.O. Culjat, M.L. Franco, C.E. Lewis, E.P. Dutson, W.S. Grundfest, J.W. Bisley, Tactile feedback induces reduced grasping force in robot-assisted surgery. IEEE Trans. Haptics **2**(2), 103–110 (2009)

55. C.-H. King, M.O. Culjat, M.L. Franco, J.W. Bisley, G.P. Carman, E.P. Dutson, W.S. Grundfest, A multielement tactile feedback system for robot-assisted minimally invasive surgery. IEEE Trans. Haptics **2**(1), 52–56 (2009)
56. M. Li, S. Luo, L. Seneviratne, T. Nanayakkara, K. Althoefer, P. Dasgupta, Haptics for multifingered palpation, in *Proceedings of the IEEE International Conference on Systems, Man, and Cybernetics* (2013), pp. 4184–4189
57. A. Bicchi, E.P. Scilingo, D. De Rossi, Haptic discrimination of softness in teleoperation: the role of the contact area spread rate. IEEE Trans. Robot. Autom. **16**(5), 496–504 (2000)
58. J.C. Gwilliam, A. Degirmenci, M. Bianchi, A.M. Okamura, Design and control of an air-jet lump display, in *Proceedings of the IEEE Haptics Symposium* (2012), pp. 45–49
59. C. Pacchierotti, L. Meli, F. Chinello, M. Malvezzi, D. Prattichizzo, Cutaneous haptic feedback to ensure the stability of robotic teleoperation systems. Int. J. Robot. Res. (2015), http://ijr.sagepub.com/content/early/2015/10/15/0278364915603135.abstract (in press)
60. C. Pacchierotti, D. Prattichizzo, K.J. Kuchenbecker, Displaying sensed tactile cues with a fingertip haptic device. IEEE Trans. Haptics in press (2015)
61. C. Pacchierotti, D. Prattichizzo, K.J. Kuchenbecker, A data-driven approach to remote tactile interaction: from a biotac sensor to any fingertip cutaneous device, in Haptics: Neuroscience, Devices, Modeling, and Applications. Eurohaptics 2014, Lecture Notes in Computer Science, Versailles, France (2014), pp. 418–424
62. C. Pacchierotti, D. Prattichizzo, K.J. Kuchenbecker, Cutaneous feedback of fingertip deformation and vibration for palpation in robotic surgery. IEEE Trans. Biomed. Eng. in press (2015)
63. M. Nakao, T. Kuroda, M. Komori, H. Oyama, Evaluation and user study of haptic simulator for learning palpation in cardiovascular surgery, in *Proceedings of the International Conference on Artificial Reality and Telexistence* (2003)
64. M.A. Giamberardino, L. Vecchiet, Pathophysiology of visceral pain. Curr. Pain Headache Rep. **1**(1), 23–33 (1997)
65. A.P. Baranowski, P. Abrams, M. Fall, *Urogenital Pain in Clinical Practice* (CRC Press, Boca Raton, 2013)
66. S.B. Schorr, Z.F. Quek, R.Y. Romano, I. Nisky, W.R. Provancher, A.M. Okamura, Sensory substitution via cutaneous skin stretch feedback, in *Proceedings of the IEEE International Conference on Robotics and Automation* (2013), pp. 2341–2346
67. T.E. Murphy, R.J. Webster, A.M. Okamura, Design and performance of a two-dimensional tactile slip display, in *Proceedings of the Eurohaptics* (2004), pp. 130–137
68. E.P. Westebring-Van der Putten, J.J. van den Dobbelsteen, R.H.M. Goossens, J.J. Jakimowicz, J. Dankelman, The effect of augmented feedback on grasp force in laparoscopic grasp control. IEEE Trans. Haptics **3**(4), 280–291 (2010)
69. D.A. Kontarinis, R.D. Howe, Tactile display of vibratory information in teleoperation and virtual environments. Presence **4**(4), 387–402 (1995)

Chapter 2
Needle Insertion in Simulated Soft Tissue

Abstract This chapter presents the very first application of the cutaneous-only sensory subtraction approach in teleoperation. It considers a simulated needle insertion in soft tissue along one direction. Part of the needle workspace is protected by a forbidden-region stiff active constraint, which is a common scenario for biopsies, deep brain stimulation and functional neurosurgery. Subjects are required to insert the needle inside the simulated soft tissue and stop the motion of the tool as soon as the presence of the stiff constraint is felt. The motion of the needle is controlled through an Omega 3 haptic interface. Accordingly to the sensory subtraction approach, the haptic feedback provided by the Omega 3 is substituted with cutaneous feedback provided by a pair of ungrounded fingertip cutaneous devices. Experiments show that the proposed cutaneous-only feedback approach, other than being intrinsically stable, improves teleoperation performance with respect to other sensory substitution techniques, such as the one using visual feedback in substitution of haptic feedback.

2.1 Introduction

Force feedback is helpful during needle advancement to detect local mechanical properties of the tissue and to distinguish between expected and abnormal resistance due, for example, to the unexpected presence of vessels or to the action of active constraints. Active constraints are software functions used in assistive robotic systems to regulate the motion of surgical tools. The motion of the surgical tool, the needle in our case, is still controlled by the surgeon, but the system constantly monitors its motion and takes some actions if it fails to follow a predetermined procedure. Active constraints play two main roles: they can either guide the motion of the tool or strictly forbid the surgeon from reaching certain regions [1]. A guiding active

---

This chapter is reprinted with kind permission from IEEE, originally published in [2].

C. Pacchierotti, *Cutaneous Haptic Feedback in Robotic Teleoperation*,
Springer Series on Touch and Haptic Systems, DOI 10.1007/978-3-319-25457-9_2

constraint attenuates the motion of the surgical tool in some predefined directions to encourage the surgeon to conform to the procedure plan. A forbidden-region active constraint seeks to prevent the needle from entering a specific region of the workspace. Forbidden-region active constraints may be introduced to protect areas that must be avoided to prevent damage of tissue and of its functionality. This is the case, for instance, of brain surgery, in which tissue manipulation in certain areas can cause serious injury to patients.

In this work we consider a simulated needle insertion in soft tissue along one direction. Part of the needle workspace is protected by a forbidden-region stiff active constraint. This is a common scenario for biopsies, deep brain stimulation and functional neurosurgery [3, 4]. Subjects are required to insert the needle inside the simulated soft tissue and stop the motion of the tool as soon as the presence of the stiff constraint is felt. The motion of the needle is controlled through an Omega 3 haptic interface. We propose to substitute the haptic feedback provided by the Omega 3 with cutaneous feedback provided by a pair of ungrounded fingertip cutaneous devices.

Section 2.2 gives a brief description of the cutaneous device used for the experiment. Sections 2.3 and 2.4 present and discuss the experimental evaluation, respectively. Finally, Sect. 2.5 addresses concluding remarks and perspectives of the work, together with its relevance for the other applications presented in the book.

2.2 An Ungrounded Fingertip Cutaneous Device

The 3-DoF ungrounded cutaneous device used in these experiments is shown in Fig. 2.1a. It consists of two platforms: one is located on the back of the finger, supporting three small DC motors, and the other one is in contact with the volar surface of the fingertip. The motors shorten and lengthen three cables to move the platform toward the user's fingertip and re-angle it to simulate contacts with arbitrarily oriented surfaces. The direction and amount of the force reflected to the user is changed

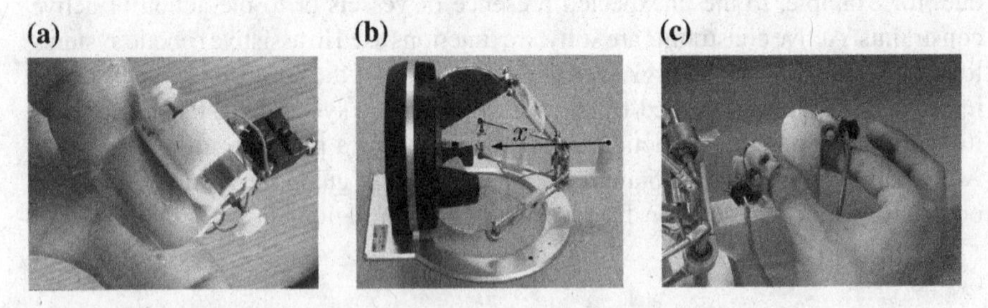

Fig. 2.1 Experimental setup. Subjects were asked to wear two cutaneous devices on the *right hand* and grasp the Omega 3 handle as shown in Fig. 2.1c. Subjects were also asked to wear two additional cutaneous devices on the thumb and index finger of the *left hand*. The motion of the Omega 3 haptic device was limited along its x–axis by three clamps. **a** The cutaneous device. **b** Omega 3 grounded device. **c** The custom handle used by a subject during the insertion

by properly controlling the cable lengths. This device is an improved version of the display presented by Minamizawa et al. [5]. In particular, the improvement consists of using three motors and a 3-DoF parallel manipulator architecture to render forces at the finger pad [6]. This device can be also seen as a simplified version of the device presented by Prattichizzo et al. [7]. However, when we ran the experiments reported in this chapter, the ungrounded device of [7] was still in a very early stage of development.

Although this device is capable of orienting and translating the mobile platform in three-dimensional space, in this work we used it as a 1-DoF system (all motors pulled the cables together), so that only forces in the sagittal plane of the finger were actuated, roughly normal to the longitudinal axis of the distal phalanx.

2.3 Experimental Evaluation

2.3.1 Experimental Setup

The experimental setup is shown in Fig. 2.1. It is composed of an Omega 3 haptic interface and four prototypes of the cutaneous device presented in Sect. 2.2. Three clamps are applied to the Omega interface to reduce its degrees of freedom from three to one (the x axis in Fig. 2.1b). A custom plastic handle is attached to the Omega's end-effector to allow the subject to use the device with two fingers (see Fig. 2.1c). The Omega 3 is used as a haptic device of the impedance type: the position of the needle, moved by the human subject, is measured, and a force signal is fed back to the user through the actuation system. The force signal accounts for either the remote contact interaction of a slave robot, in a classical teleoperation scenario, or of the virtual environment, as in our simulated scenario. Subjects were asked to wear two cutaneous devices on the right hand, on the thumb and index fingers, and to use the handle as shown in Fig. 2.1c. To investigate the role of feedback localization with respect to the hand involved in the task, subjects were also asked to wear two additional cutaneous devices on the thumb and index fingers of the contralateral hand.

The haptic handle teleoperates the needle in a virtual environment simulating the insertion in soft tissue with a stiff constraint. The needle moves along a single axis as shown in Fig. 2.2, where the needle, the tissue surface, and the stiff constraint are shown. The stiff constraint and the portion of the needle inside the tissue surface are not shown to the subject. The contact force between the needle and the tissue is calculated according to a visco-elastic model. The subject steering the needle feels a resistive force while penetrating the tissue, and an opposite force while trying to pull the needle out. In real scenarios, these forces are either measured from force sensors or estimated from other parameters. In this work a simple simulation of the soft tissue is used. The aim of this work is not the design an accurate tissue simulator based, for example, on FEM techniques [8], but to validate the proposed sensory subtraction approach.

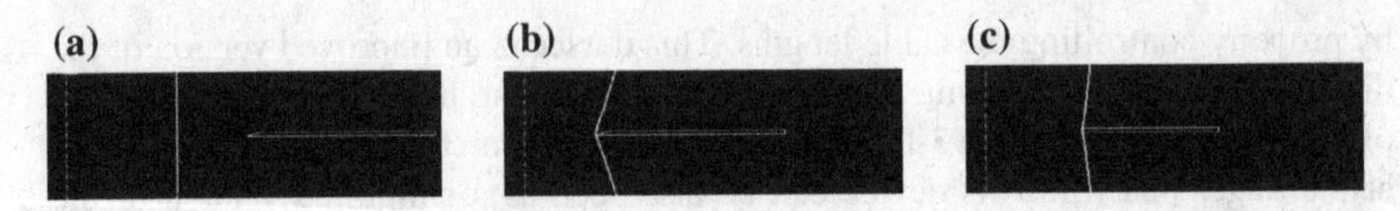

Fig. 2.2 The virtual environment is composed of the needle (*white*), driven by the operator, the deformable tissue (*cyan*), and the stiff constraint (*red*). The position of the needle x_n is linked to the position of the haptic device end-effector. The stiff constraint and the portion of the needle inside the tissue surface are not shown to the subject. **a** No needle-tissue contact. **b** Needle reaches the tissue. **c** Needle penetrates the tissue (color figure online)

A spring $k_t = 2\,\text{N/m}$ and a damper $b_t = 5\,\text{Ns/m}$ are used to model the contact force f_t between the needle and the tissue, while a spring $k_{sc} = 3000\,\text{N/m}$ is used to model the contact force f_{sc} between the needle and the stiff constraint. For the sake of simplicity, we assume that the mass of the tissue $m_t = 1$ kg is concentrated at the contact point. The viscous coefficient of the body beneath the tissue is $v_t = 0.7\,\text{Ns/m}$.

As for the haptic rendering, the interaction is designed according to the god-object model [9] and the position of the Omega handle is linked to the needle position x_n moving in the virtual environment. The initial position of the surface of the tissue is set to $\bar{x}_t = 20\,\text{mm}$ and the stiff constraint is located at $\bar{x}_{sc}$. Tissue position x_t changes according to the interaction with the needle, which is able to penetrate the surface only when the contact force f_h is larger than a predetermined threshold ($f_p = 0.1\,\text{N}$). To extend the workspace in the virtual environment, we introduce a scale factor of 3 between the position of the needle and the subject's hand.

It is thus possible to discriminate four different operating conditions for the needle-environment interaction model here presented:

- no contact (see Fig. 2.2a),
- contact without penetration (see Fig. 2.2b),
- penetration within the safe area (see Fig. 2.2c), and
- penetration and contact with the stiff constraint.

In the first case, since the needle is out of the tissue, the model is designed to feed back no force to the subject and the surface of the tissue tends to return to its predetermined initial position $\bar{x}_t$. The dynamics of the interaction for the no contact case is therefore

$$\begin{cases} m_t\,\ddot{x}_t = -k_t\,(x_t - \bar{x}_t) - b_t\,\dot{x}_t, \\ f_h = 0. \end{cases}$$

When the needle touches the tissue, but the force f_h is not yet sufficient to penetrate it, the tissue surface is deformed by the movement of the needle. In this case, the dynamic model and the contact force to be fed back to the subject are

$$\begin{cases} x_t = x_n, \\ f_h = -k_t\,(x_t - \bar{x}_t) - b_t\,\dot{x}_t. \end{cases}$$

As soon as $f_h > f_p$, the needle penetrates the surface and while the needle is inside the tissue, the dynamics and the contact force are computed as

$$\begin{cases} m_t\,\ddot{x}_t = -k_t\,(x_t - \bar{x}_t) - b_t\,\dot{x}_t - \upsilon_t\,(\dot{x}_t - \dot{x}_n), \\ f_h = -\upsilon_t\,(\dot{x}_t - \dot{x}_n). \end{cases}$$

If the subject steers the needle toward the unsafe workspace area delimited by the stiff constraint, a force will be fed back to the subject in order to avoid the penetration of the needle in the forbidden area:

$$f_{sc} = -k_{sc}\,(x_n - \bar{x}_{sc}).$$

The haptic device measures the position of the subject's hand, sends it to the controller and then the virtual environment computes the force feedback and the dynamics of the tissue. The controller then sends the force back to the user through either the haptic device or the substitutive (cutaneous or visual) condition, as detailed in the next section.

2.3.2 Design of the Experiments

Four alternative feedback conditions were compared in the experiments: (full) *haptic* feedback, applied by the actuators of the haptic interface, *visual* feedback in substitution of haptic feedback, or *cutaneous* feedback in substitution of haptic feedback, applied by the ungrounded devices either on the fingers holding the handle or on the fingers of the contralateral hand. The substitutive visual feedback consisted in showing a horizontal bar depicting the contact force f_h registered at the needlepoint.

Subjects were asked to wear the four cutaneous devices for the whole duration of the experiments, and to grasp the handle with their right hand as shown in Fig. 2.1c. The subject's hand was positioned with its longitudinal axis at 90° from the Omega x-axis. The position of the subject's hand with respect to the joystick was checked before the beginning of each experiment. To prevent changes in the perceived direction of the feedback force generated by the Omega 3, subjects were instructed to move the forearm rather than the wrist while moving the device. During the experiments, the subjects maintained the initial orientation of the fingers with respect to the handle, which was the only natural way of grasping the handle for the 1-DoF task.[1]

[1] A modification of the way the fingers grasp the handle would imply that the perceived direction of the feedback force changes if haptic feedback is used, whereas it would not change with cutaneous-only feedback. This issue must be considered while trying to extend the sensory subtraction paradigm to multi-DoF tasks, since the results may be affected by this change of direction of the force vector. Thus, the position of the subject's hand with respect to the input device must be carefully monitored before and during the experiments.

The task consisted in inserting the needle into the soft tissue and stopping its motion as soon as the presence of the stiff constraint was perceived. After 5 s of continuous contact with the constraint, the system played a sound beep. Subjects were instructed to pull the needle out of the tissue as soon as the sound was heard. In all the considered conditions, regardless of particular feedback condition employed, visual feedback on needle insertion was provided to the subjects, showing the portion of the needle outside the tissue and the surface of the tissue. The stiff constraint and the portion of the needle inside the tissue were not visible (see Fig. 2.2). No information on the feedback conditions was provided, neither on their nature (except from visual feedback in substitution of force feedback) nor on the particular order with which they were going to be presented to the subject. Both the sequence of the feedback conditions and the initial position of the stiff constraint were randomized.

We carried out three experiments:

- *experiment #1*: twenty-four repetitions of the needle insertion task described above,
- *experiment #2*: two additional repetitions of the needle insertion task, during which the position of the stiff constraint was changed suddenly and unexpectedly,
- *experiment #3*: same as experiment #1, but in presence of a time delay in the haptic loop.

The first experiment aimed at demonstrating that (1) there is no relevant degradation of performance in the task when haptic feedback is substituted with cutaneous feedback, and that (2) using cutaneous-only feedback leads to better performances with respect to visual feedback in substitution of force feedback. Moreover, the experiment investigated if the localization of the cutaneous feedback on the fingertips responsible for handling the needle is an important factor.

The second experiment aimed at showing that using the cutaneous devices prevents the handle (and so the needle) from moving in unwanted directions in case of sudden and unpredictable changes of the position of the stiff constraint.

The third experiment aimed at confirming the well known result that there are no instability behaviors, not even in presence of delays, while using cutaneous feedback devices.

2.3.3 Experiment #1: Comparison of the Feedback Conditions

Sixteen participants (13 males, 3 females, age range 21–28) took part in the experiment, all of whom were right-handed. Eight of them had previous experience with haptic interfaces. None of the participants reported any deficiencies in the perception abilities. Each participant made 24 repetitions of the needle insertion task, with six randomized trials for each feedback condition:

- *visual feedback* by the moving horizontal bar (condition VF);
- *haptic feedback* by the grounded haptic interface (condition HF);

- *cutaneous feedback* by the ungrounded devices, provided to the hand holding the handle (condition CF);
- *cutaneous feedback* by the ungrounded devices, provided to the contralateral hand (condition CCF).

The experiment lasted 9.13 min on average, including the two additional repetitions of experiment #2, which followed the twelfth and the 24th repetitions of experiment #1 (see Sect. 2.3.4 for details). Subjects thus performed a total of 26 trials, whereof 24 were included in the results of experiment #1.

With the aim of comparing the different feedback conditions, we recorded the position x_n of the needle, steered by the subject's hand, and the penetration inside the stiff constraint $p = \bar{x}_{sc} - x_n$. The average penetration $\bar{p}$ and the maximum penetration $\bar{p}_M$ were analyzed. Data resulting from different trials of the same condition, performed by the same subject, were averaged before comparison with other conditions. Such values provide a measure of accuracy (average penetration) and overshoot (maximum penetration) in reaching the target depth. A null value in both metrics denotes the best performance, while a positive value indicates that the subject overran the target. Both measures can be considered particularly relevant to the surgical task, as an excessive penetration of the needle can result in permanent damage of tissues.

Figure 2.3 shows the positions of the needle (red patch) and of the tissue surface (green patch) versus time. The time of different trials was synchronized at the time the needle first enters the constraint ($t = 0$), while positions were divided by the depth of the stiff constraint, which varied randomly among trials, and are presented as percentage. Trajectories were averaged among subjects for each feedback condition, and average trajectories plus/minus standard deviations are shown. The position of the stiff constraint (dashed red line, 100 %) and the initial position of tissue surface (dashed green line, 0 %) are shown as well. The black lines represent the instants when the average trajectory enters the tissue (left line) and when the sound beep is played (right line).

Figures 2.4 and 2.5 show, respectively, the average and maximum penetrations beyond the stiff constraint for each feedback condition (means and standard deviations are plotted). All column data passed the D'Agostino-Pearson omnibus K2 normality test. Comparison of the means among the feedback conditions was tested using one-way repeated measures analysis of variance (ANOVA). The means of average penetration (Fig. 2.4, $F_{3,45} = 106.5$, $p < 0.001$) and the means of maximum penetration (Fig. 2.5, $F_{3,45} = 81.89$, $p < 0.001$) differed significantly among the feedback conditions. Posthoc analyses (Bonferroni's multiple comparison test) revealed statistically significant difference between all pairs of conditions, both in terms of average penetration (Fig. 2.4, $p < 0.001$ for all pairs) and in terms of maximum penetration of the needle (Fig. 2.5, $p < 0.05$ for CF vs. CCF, and $p < 0.001$ for all other pairs). Results indicate that the proposed sensory subtraction condition (CF) yields an intermediate performance between haptic feedback (HF, best performance) and visual feedback (VF, worst performance), in terms of both average and maximum penetration beyond the stiff constraint. These results demonstrate also

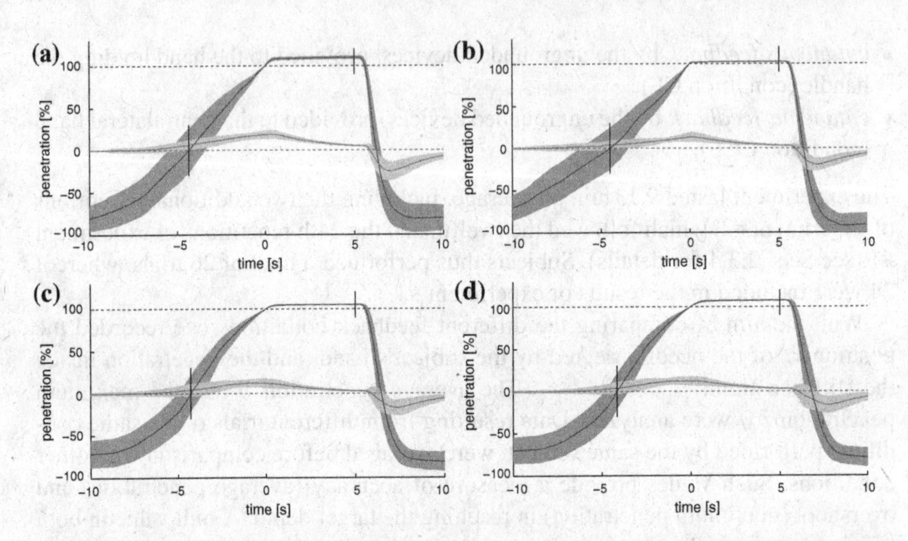

Fig. 2.3 Penetration of the needle (*red patch*) and position of the tissue surface (*green patch*) versus time for experiment #1. Average trajectories among subjects and their standard deviations are plotted. The position of the stiff constraint (*dashed red line*) and the initial position of tissue surface (*dashed green line*) are shown as well. The *black lines* represent the instants when the average trajectory enters the tissue (*left line*) and when the sound beep is played (*right line*). **a** Visual feedback (VF). **b** Haptic feedback (HF). **c** Cutaneous feedback on the hand holding the handle (CF). **d** Cutaneous feedback on the contralateral hand (CCF) (color figure online)

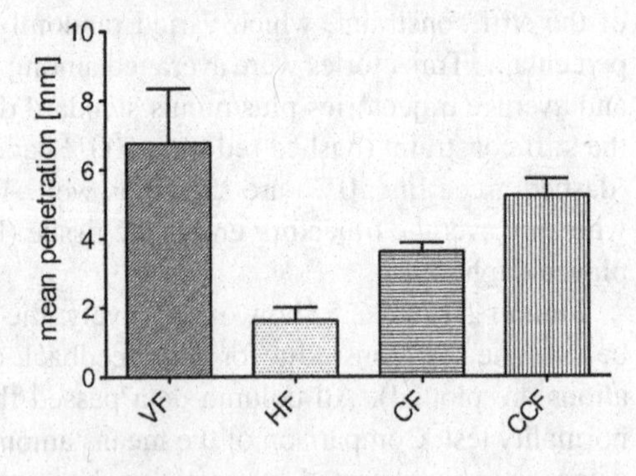

Fig. 2.4 Experiment #1: average penetration beyond the stiff constraint (mean and SD), for the visual (VF), haptic (HF) and cutaneous feedback modes (CF, CCF). A null value of this metric indicates high accuracy in reaching the target depth

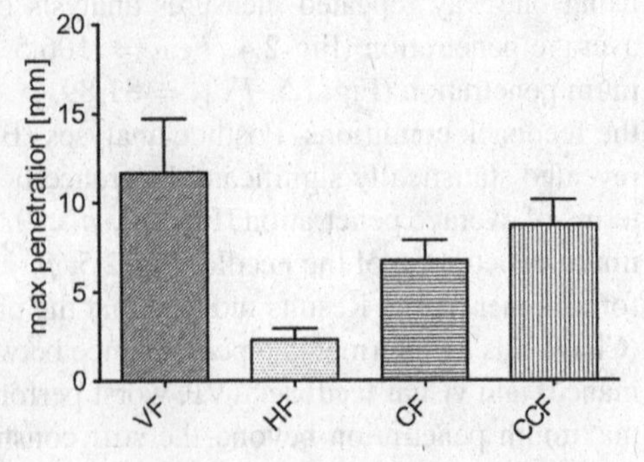

Fig. 2.5 Experiment #1: maximum penetration beyond the stiff constraint (mean and SD), for the visual (VF), haptic (HF) and cutaneous feedback modes (CF, CCF). A null value of this metric indicates no overshoot in reaching the target depth

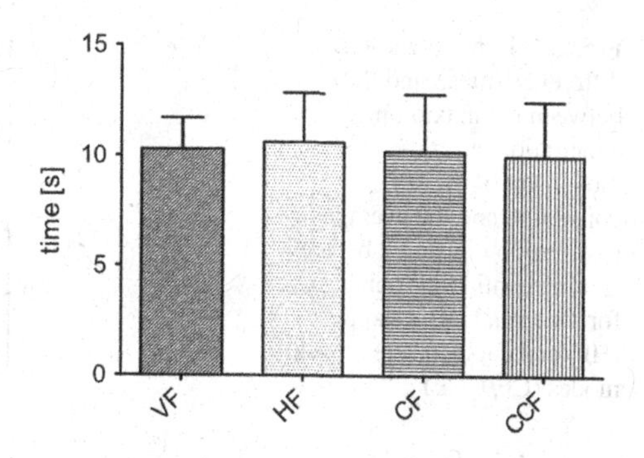

Fig. 2.6 Experiment #1: time elapsed (mean and SD) between the first contact with the tissue and the sound played after 5 s of continuous contact with the stiff constraint, for the visual (VF), haptic (HF) and cutaneous feedback modes (CF, CCF)

that the cutaneous devices provide a more reliable form of feedback if applied to the fingertips which are responsible for holding the end-effector (CF) with respect to contralateral hand stimulation (CCF), suggesting that the localization of cutaneous feedback is crucial in this setting. Nonetheless, cutaneous feedback is more effective than visual feedback (VF) even when it is applied to the contralateral hand (CCF), indicating that not only the localization but also the nature of the sensation provided is relevant to the performance of the task.

Figure 2.6 shows the average time elapsed between the instant the needle penetrates the tissue and the instant it reaches 5 s of continuous contact with the stiff constraint. Column data failed to pass the normality test, so we used the Friedman non-parametric test to analyze variance. Results indicate that there is no statistically significant difference between the feedback conditions in this metric. We may read this result by saying that the subjects became equally confident with all the feedback conditions proposed.

2.3.4 Experiment #2: Sudden Change of the Stiff Constraint

This second experiment evaluated the effect on needle position of a sudden and unpredictable change of the position of the stiff constraint, in the presence of the four feedback conditions described before (visual, haptic and the two cutaneous). The needle insertion task was the same as that described in Sect. 2.3.3. However, after 5 s of continuous contact, the depth of the stiff constraint was increased unexpectedly, so that the virtual environment suddenly fed back no force to the user. At the same time, the sound beep was produced as in the other repetitions of the needle insertion task, signaling the subject to extract the needle.

The test was performed during two additional repetitions of experiment #1. To ensure the surprise effect, each subject performed only two additional repetitions (using two different feedback conditions). A total of 32 trials were thus recorded for experiment #2: 8 trials for each feedback condition, performed by eight different subjects. The first additional trial was run after the 12th trial of experiment #1, the second after the 24th. No information was provided to the subjects about the

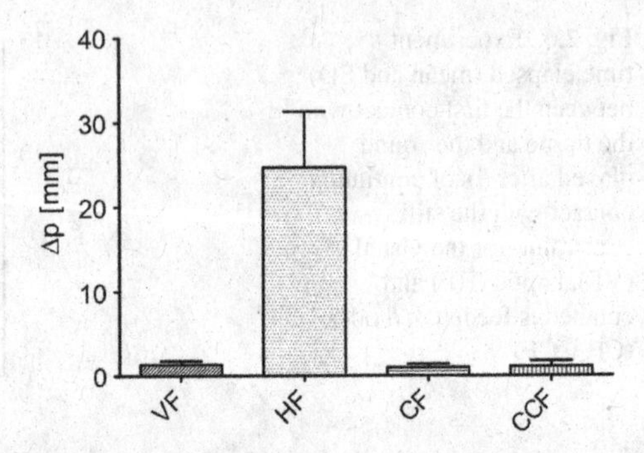

Fig. 2.7 Experiment #2: difference (mean and SD) between the maximum penetration, after the movement of the stiff constraint, and the average penetration registered before (during continuous contact), for the visual (VF), haptic (HF) and cutaneous feedback modes (CF, CCF)

additional trials, which followed immediately the previous ones. A 30 s rest was given to all subjects after the first additional trial, before continuing with the second part of experiment #1. Subjects did not know that the position of the stiff constraint was going to change and that they were performing a different task with respect to the others.

Figure 2.7 shows the difference Δp between the maximum penetration registered after the perturbation and the average penetration observed in the 5 s before (continuous contact). All column data passed the D'Agostino-Pearson omnibus K2 normality test. Comparison of the means among the feedback conditions was tested using one-way ANOVA (no repeated measures). The means differed significantly among the feedback conditions ($F_{3,28} = 100.3$, $p < 0.001$). Posthoc analyses (Bonferroni's multiple comparison test) revealed statistically significant difference between haptic feedback (HF) and each alternate condition (VF, CF, CCF, $p < 0.001$). Results indicate that the presence of kinesthetic feedback may induce significantly greater unwanted motions of the needle with respect to the three non-kinesthetic feedback modes used in the experiments (the visual and the two cutaneous-only conditions). In fact, when the constraint moves during the haptic condition (HF), the subject's arm is counteracting an external force which suddenly drops.

Figure 2.8 shows the positions of the needle (red patch) and of the tissue surface (green patch) versus time for experiment #2. Data were synchronized, normalized and averaged among subjects as for the charts of Fig. 2.3.

2.3.5 *Experiment #3: Stability with Time Delay*

As other sensory substitution techniques, the main advantage of the proposed sensory subtraction approach is that it makes the haptic loop intrinsically stable. No instability behaviors occur, even in presence of large delays.

To support this hypothesis, we carried out a third experiment. We used the same protocol of experiment #1, including the feedback conditions and number of repetitions (24) per subject, but we introduced a delay of 50 ms in the haptic loop, between

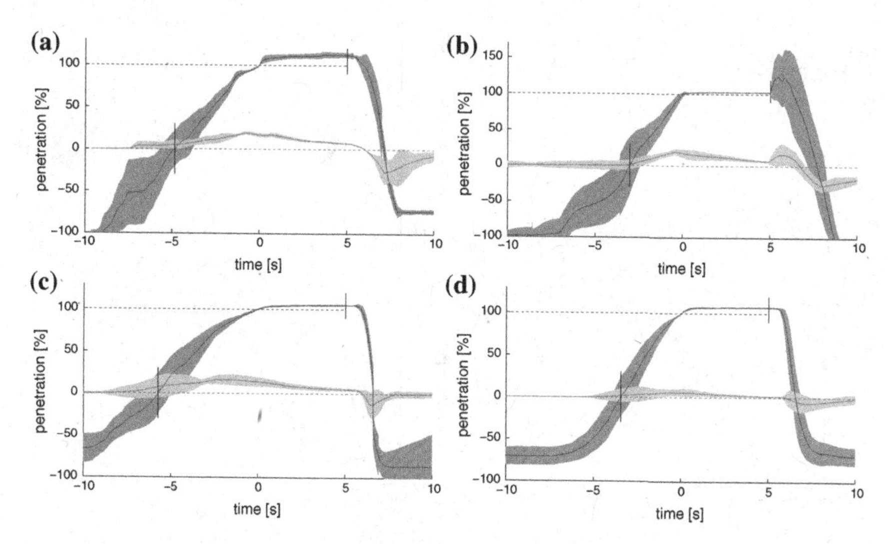

Fig. 2.8 Penetration of the needle (*red patch*) and position of the tissue surface (*green patch*) versus time for experiment #2, with the stiff constraint suddenly removed after 5 s of continuous contact. Average trajectories among subjects and their standard deviations are plotted. The position of the stiff constraint (*dashed red line*) and the initial position of tissue surface (*dashed green line*) are shown as well. The *black lines* represent the instants when the average trajectory enters the tissue (*left line*) and when the stiff constraint is removed and the sound beep is played (*right line*). **a** Visual feedback (VF). **b** Full haptic feedback (HF). **c** Cutaneous feedback on the hand holding the handle (CF). **d** Cutaneous feedback on the contralateral hand (CCF) (color figure online)

the virtual environment and either the haptic handle, the cutaneous devices or the visual rendering of force. Recent literature denotes the relevance of delays in teleoperated surgical tasks [10]. It is worth noting that instability of haptic feedback in the presence of time delays can be addressed by designing appropriate control systems [11–13]. Nonetheless, to emphasize the intrinsic stability of cutaneous feedback, this method was not used in the trials.

Ten participants (8 males, 2 females, age range 20–26) took part in the experiment, all of whom were right-handed and five of whom had previous experience with haptic interfaces. None of the participants reported any deficiencies in the perception abilities. The experiment lasted 8.39 min on average.

Figure 2.9 shows the positions of the needle (red patch) and of the tissue surface (green patch) versus time for experiment #3. A video of a representative run showing the instability issue is available as supplemental material at http://extras.springer.com/978-3-319-25455-5 and at http://goo.gl/Zf0QAB. Data were synchronized, normalized and averaged among subjects as for Fig. 2.3. By comparing the charts with those in Fig. 2.3, we can notice that the instability occurred only with haptic feedback, i.e., only in the presence of the kinesthetic part of the interaction. Significant oscillations of the needle are likely to bring not only a greater penetration of the needle in the stiff constraint, but also a longer task completion time.

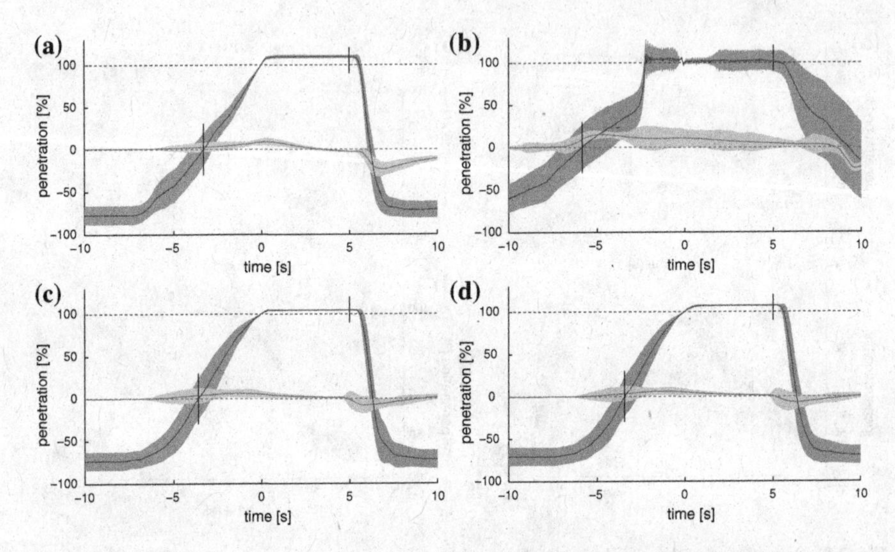

Fig. 2.9 Penetration of the needle (*red patch*) and position of tissue surface (*green patch*) versus time for experiment #3, with a 50 ms network delay in the haptic loop. Average trajectories among subjects and their standard deviations are plotted. The position of the stiff constraint (*dashed red line*) and the initial position of tissue surface (*dashed green line*) are shown as well. The *black lines* represent the instants when the average trajectory enters the tissue (*left line*) and when the sound beep is played (*right line*). **a** Visual feedback (VF). **b** Full haptic feedback (HF). **c** Cutaneous feedback on the hand holding the handle (CF). **d** Cutaneous feedback on the contralateral hand (CCF) (color figure online)

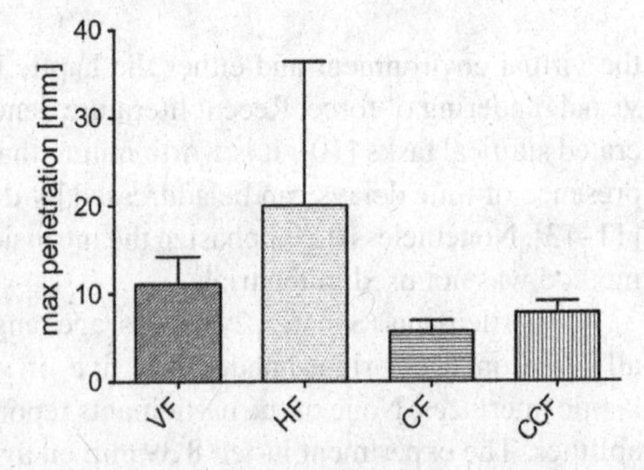

Fig. 2.10 Experiment #3: maximum penetration beyond the stiff constraint (mean and SD), for the the visual (VF), haptic (HF) and cutaneous feedback modes (CF, CCF), with a 50 ms network delay in the loop

Figure 2.10 shows the maximum penetration beyond the constraint in the presence of the delay. Haptic feedback group data (HF) failed to pass the normality test, so the Friedman non-parametric test was used to analyze variance. The test indicated statistically significant difference between the feedback conditions ($p < 0.001$). Posthoc analyses (Dunn's multiple comparison test) revealed statistically significant difference between haptic feedback (HF) and both cutaneous conditions (CF, $p < 0.001$; CCF, $p < 0.05$), and between cutaneous feedback (CF) and visual feedback (VF, $p < 0.001$). Results indicate that the subjects, while receiving the

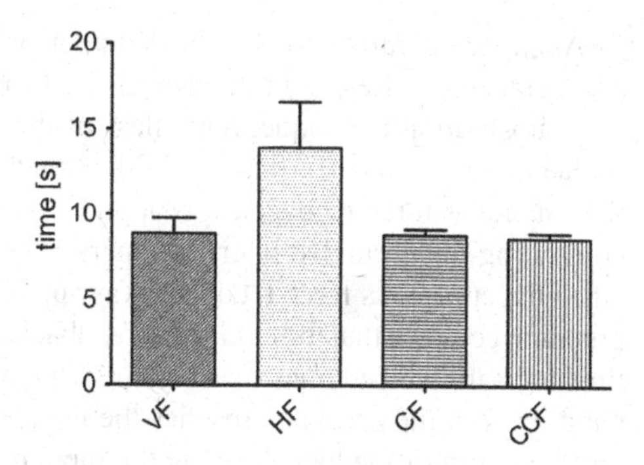

Fig. 2.11 Experiment #3: time elapsed between the first contact with the tissue and the sound beep, for the visual (VF), haptic (HF) and cutaneous feedback modes (CF, CCF), with a 50 ms network delay in the haptic loop

full haptic feedback in the presence of a time delay, reached a significantly greater peak penetration in the stiff constraint with respect to that obtained while receiving feedback from the ungrounded cutaneous devices, regardless of the localization of cutaneous feedback. The same result was obtained when subjects received visual feedback of force instead of cutaneous feedback on the fingers which are responsible for handling the needle.

Figure 2.11 shows, for each feedback condition and in the presence of the time delay, the mean time elapsed between the first penetration in the tissue and the instant the needle reaches 5 s of stable contact with the stiff constraint. Haptic feedback and cutaneous feedback group data failed to pass the normality test, so the Friedman non-parametric test was used to analyze variance. The test indicated statistically significant difference between the feedback conditions ($p < 0.001$). Posthoc analyses (Dunn's multiple comparison test) revealed statistically significant difference between haptic feedback (HF) and all other feedback conditions (VF, $p < 0.01$; CF, $p < 0.05$; CCF, $p < 0.001$). Results indicate that the time needed to complete the task was significantly greater while receiving the kinesthetic feedback with respect to the other non-kinesthetic feedback conditions. Such a difference had not been observed in the absence of time delays (see Fig. 2.6), and must be related to instability.

2.4 Discussion

The first experiment evaluated the effectiveness of the sensory subtraction technique. The results of this experiment indicate that the subjects, while receiving visual feedback (VF) in substitution of force feedback, reached a significantly greater average and maximum penetration in the stiff constraint (worst performance) in comparison with that obtained while receiving either full haptic (HF) or cutaneous-only feedback (CF and CCF). Conditions CF and CCF provided intermediate performance between visual and haptic feedback. No difference between groups was observed in terms of task completion time.

As expected, full haptic feedback outperformed all the other feedback conditions. The cutaneous-only condition proved itself to be more intuitive than the other sensory substitution technique, regardless of the localization of the cutaneous devices (either on the hand performing the task or on the contralateral hand). However, when the cutaneous force feedback was applied to the contralateral hand (i.e., the one not controlling the grounded interface), performance degraded with respect to the case when the cutaneous force feedback was applied to the active hand. A possible interpretation could be that the cutaneous feedback applied to the contralateral hand needs time for transcallosal transmission to reach the hemisphere controlling the operating hand. In fact, the feedback reaches the hemisphere of the brain not involved in the motor control of the hand moving the input device, and for this reason may require more time to be transformed in motor action [14].

These results suggest that the sensory subtraction approach can be successfully used in substitution of traditional haptic feedback, with a significant improvement of performance with respect to visual sensory substitution techniques. Not only the type of feedback, cutaneous rather than visual, but also the place where it is applied is important. The best performance is in fact obtained when the cutaneous devices are worn on the hand involved in the task. This result can be explained by considering that the area of application of the force and the particular design of the cutaneous devices proposed provide the user with a direct and intuitive measure of the contact force being substituted, thus producing a more natural interaction with the device.

However, during the first experiment, performance degraded when kinesthetic feedback information was removed (HF vs. CF). One possible explanation is that kinesthetic force played a role in arm dynamics during the execution of the needle insertion task. In particular, the virtual force helped the subject in stopping hand motion when the stiff constraint was reached, which is the main reason for using stiff constraints indeed. Conversely, in the proposed cutaneous-only condition, no physical aid is provided to the user to accomplish the task, so any arm motion derives entirely from motor control. The resulting benefit is that unwanted motions can be drastically reduced in critical situations, as shown in experiment #2. Nonetheless, without adequate sense of touch, achieving normal and top performance in tasks that require high levels of dexterity is extremely difficult [15]. Moreover, even simple tactile information can be effective, both in virtual and in real environments. For example, major gains in body posture control in real environments can be obtained from minimal touch information applied to a fingertip [16]. This may explain why cutaneous only tasks were better executed than the substituted visual condition tasks (VF).

One major advantage of sensory subtraction is that, despite the fact that the interaction is closer to haptic rendering, no unwanted movements are likely to be produced during the execution of guided tasks. This achievement, that is corroborated by the results of experiment #2, is crucial in applications where safety is paramount, such as robot-aided surgery, in which unwanted movements of the surgeon's hand induced by knesthetic feedback may produce serious damages to the patient. The absence of unwanted movements, even in the case of sudden and unpredictable changes of the position of the stiff constraint, can be explained by considering that, in the cutaneous feedback conditions, kinesthetic feedback is completely missing, so that subjects can maintain a stable contact with the stiff constraint without exerting an active force on the handle.

The last experiment showed that, in the presence of a transmission delay, full haptic feedback can bring the system near to instability, as significant oscillations of needle position occurred, whereas cutaneous (and visual) feedback allows a stable contact with the stiff constraint surface. The occurrence of instability with a relatively small time delay may be due to the particular (stiff) setting of the experimental device used in the experiments. However, the fact that kinesthesia can bring instability in haptic teleoperation in the presence of time delays is well-know in the literature of haptics and teleoperation.

Another drawback of using full haptic feedback in presence of transmission delays is the longer time needed to complete the task. Statistical analysis on task completion times showed that, in case of no delay, there are no significant differences between the four different feedback conditions, while in the presence of a network delay, task completion time using haptic feedback can be significantly greater than that obtained using cutaneous-only feedback.

2.5 Conclusions

The work presented in this chapter showed how cutaneous feedback applied to the thumb and index finger pads can be effectively used to substitute full haptic feedback in robotic teleoperation. The main advantage of using cutaneous feedback is that the stability of the haptic loop is intrinsically guaranteed. This can be very convenient for critical applications such as robotic surgery. Chapter 4 will address and analyze the use of cutaneous feedback in robotic surgery systems. Note also that actuation for cutaneous displays usually requires less power and it is less bulky than that required to provide haptic feedback, with a direct effect on simplifying mechanical design and reducing costs.

The main drawback of the proposed approach is that, like for other sensory substitution techniques, the performance is lower than that of full haptic feedback. However, differently from other substitution techniques, our approach conveys force feedback exactly where it is expected and provides the subject with a direct and co-located perception of the contact force. This degradation of performance with respect to full haptic feedback approaches is the main motivation that pushed us toward solutions mixing cutaneous and kinesthetic stimuli, as I will further discuss in Chap. 5. Another weak point of the work presented in this chapter is that the experiments were carried out in a simulated 1-DoF scenario. In the next chapters we will extend our evaluation of the sensory subtraction approach to more challenging scenarios. Chapter 3 considers a remote peg-in-hole task, both in simulated and real environments. Chapter 4 presents an application of the sensory subtraction idea in a remote palpation task using the da Vinci Surgical System.

References

1. D. Prattichizzo, C. Pacchierotti, G. Rosati, Cutaneous force feedback as a sensory subtraction technique in haptics. IEEE Trans. Haptics **5**(4), 289–300 (2012)
2. J.J. Abbott, P. Marayong, A.M. Okamura, Haptic virtual fixtures for robot-assisted manipulation. Robot. Res. 49–64 (2007)
3. D. De Lorenzo, E. De Momi, I. Dyagilev, R. Manganelli, A. Formaglio, D. Prattichizzo, M. Shoham, G. Ferrigno, Force feedback in a piezoelectric linear actuator for neurosurgery. Int. J. Med. Robot. Comput. Assist. Surg. **7**(3), 268–275 (2011)
4. B. Davies, Robotic devices in surgery. Minim. Invasive Ther. Allied Technol. **12**(1), 5–13 (2003)
5. K. Minamizawa, S. Fukamachi, H. Kajimoto, N. Kawakami, S. Tachi, Gravity grabber: wearable haptic display to present virtual mass sensation, in *Proceeding of the ACM Special Interest Group on Computer Graphics and Interactive Techniques Conference* (2007), p. 8–es
6. B. Dasgupta, T.S. Mruthyunjaya, The stewart platform manipulator: a review. Mech. Mach. Theory **35**(1), 15–40 (2000)
7. D. Prattichizzo, F. Chinello, C. Pacchierotti, M. Malvezzi, Towards wearability in fingertip haptics: a 3-dof wearable device for cutaneous force feedback. IEEE Trans. Haptics **6**(4), 506–516 (2013)
8. S. Cotin, H. Delingette, N. Ayache, Real-time elastic deformations of soft tissues for surgery simulation. IEEE Trans. Vis. Comput. Gr. **5**(1), 62–73 (1999)
9. C.B. Zilles, J.K. Salisbury, A constraint-based god-object method for haptic display, in *Proceedings IEEE/RSJ International Conference on Intelligent Robots and Systems*, vol. 3 (1995), pp. 146–151
10. I. Nisky, A. Pressman, C.M. Pugh, F.A. Mussa-Ivaldi, A. Karniel, Perception and action in teleoperated needle insertion. IEEE Trans. Haptics **4**(3), 155–166 (2011)
11. G. Niemeyer, J.J.E. Slotine, Stable adaptive teleoperation. IEEE J. Ocean. Eng. **16**(1), 152–162 (1991)
12. Y. Ye, P.X. Liu, Improving haptic feedback fidelity in wave-variable-based teleoperation orientated to telemedical applications. IEEE Trans. Instrum. Measurement **58**(8), 2847–2855 (2009)
13. T.M. Lam, M. Mulder, M.M. Van Paassen, Haptic feedback in UAV tele-operation with time delay. J. Guidance, Control Dyn. **31**(6), 1728–1739 (2008)
14. E.R. Kandel, J.H. Schwartz, T.M. Jessell et al., *Principles of Neural Science* (McGraw-Hill, New York, 2000)
15. G. Robles-De-La-Torre, The importance of the sense of touch in virtual and real environments. IEEE Multimed. **13**(3), 24 (2006)
16. J.J. Jeka, P. Ribeiro, K. Oie, J.R. Lackner, The structure of somatosensory information for human postural control. Mot. Control **2**(1), 13–33 (1998)

Chapter 3
Peg-in-Hole in Simulated and Real Scenarios

Abstract This chapter presents the second application of the cutaneous-only sensory subtraction approach in robotic teleoperation. It considers a peg-in-hole task, both in virtual and real environments. The master system is composed of two novel 3-DoF fingertip cutaneous displays, attached to the end-effectors of two Omega 3 haptic interfaces. In the first experiment, subjects interact via two virtual spheres with a virtual environment composed of a peg and a board with two holes. In the second experiment, the slave system is composed of a DLR-HIT Hand II attached to a 6-DoF manipulator KUKA KR3, and the remote environment is composed of a peg and a rigid board with two holes. Accordingly to the sensory subtraction approach, the haptic feedback provided by the Omega 3 is substituted with cutaneous feedback provided by the two 3-DoF fingertip cutaneous devices. Results assessed the feasibility and effectiveness of the proposed cutaneous-only approach. Cutaneous feedback was outperformed by full haptic feedback provided by grounded haptic interfaces, but it outperformed conditions providing no force feedback at all. Moreover, cutaneous feedback always kept the system stable, even in the presence of destabilizing factors such as communication delays and hard contacts.

3.1 Introduction

The weakest point of the work presented in Chap. 2 is the experiment being carried in a simulated 1-DoF scenario. Although interesting and very promising, our sensory subtraction idea needs to be validated in a more challenging environment if we want to prove its applicability and effectiveness in teleoperation. For this reason, we designed a new set of experiments to extend the results presented in Chap. 2 to a more challenging and general scenario that considers a remote peg-in-hole task, both in simulated and real environments. Such a task requires moderately complex movement and force application, and it has been previously used to measure task performance of haptic feedback in teleoperation [1–3].

This chapter is reprinted with kind permission from SAGE, originally published in [4].

C. Pacchierotti, *Cutaneous Haptic Feedback in Robotic Teleoperation*,
Springer Series on Touch and Haptic Systems, DOI 10.1007/978-3-319-25457-9_3

Once again, we propose to substitute the haptic feedback provided by grounded haptic interfaces with cutaneous feedback provided by a pair of fingertip cutaneous devices. We expect this kind of feedback to make the teleoperation system intrinsically stable, since the cutaneous force applied does not affect the position of the master device, thus opening the feedback loop (see Chap. 1). Furthermore, as in Chap. 2, we expect the human operator to perform the given task in an intuitive way, since cutaneous stimuli provide the user with a direct and co-located perception of the feedback force. We analyzed the implications and outcomes of such an approach for two teleoperated peg-in-hole tasks, in virtual and real scenarios, employing a pair of novel 3-DoF cutaneous displays. Each device exerts cutaneous stimuli at the fingertip by applying forces to the vertices of a rigid platform.

Section 3.2 presents the cutaneous device used for the experiments. Section 3.3 presents the experimental evaluation, while Sect. 3.4 discusses the results. Finally, Sect. 3.5 addresses concluding remarks and perspectives of the work, together with its relevance for the other applications presented in the book.

3.2 A Fingertip Cutaneous Device for the Omega Interface

To demonstrate the feasibility and performance of the sensory subtraction approach, we designed a novel 3-DoF cutaneous haptic device. It is able to provide cutaneous stimuli at the fingertip, and it can be easily attached to the end-effector of commercial haptic interfaces. Figure 3.1a sketches the main idea, while a prototype of the device is shown in Fig. 3.1b. A video of the device is available as supplemental material

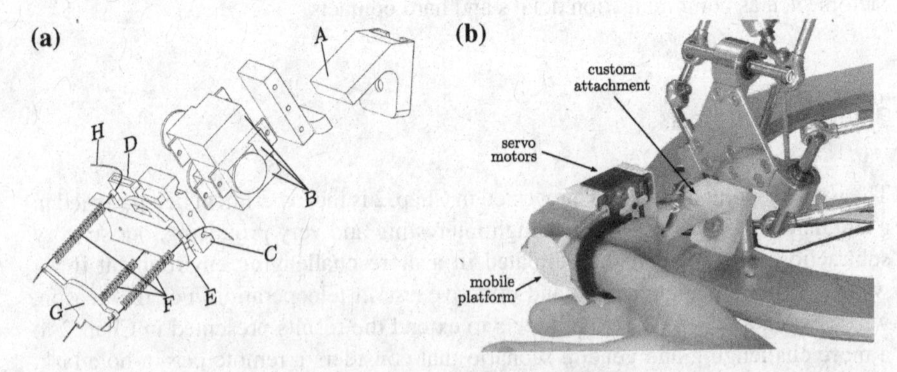

Fig. 3.1 The 3-DoF fingertip cutaneous device. It is composed of a static body (*A*, *C*–*E*) that houses three servo motors (*B*) and a mobile platform (*G*) that applies the requested stimuli to the fingertip. Three cables (*H*) and springs (*F*) connect the two platforms. By controlling the cable lengths, the motors can orient and translate the mobile platform in three-dimensional space. The device fastens to the finger with a fabric strap fixed to part *D*. The device can be easily attached to the end-effector of different commercial haptic devices

Table 3.1 Technical specifications of the fingertip cutaneous device

General	
Actuators	HS-55 servo motor (max current 1 A)
Power Supply	External adapter 6 V 1 A
Controller	Atmega 328 on an Arduino Nano board
Motor bandwidth	100 Hz
Platform	
Max normal force	20 N
Max tangential force	8 N
Max roll angle	25°
Max pitch angle	25°
Max speed	21 mm/s

at http://extras.springer.com/978-3-319-25455-5 and at http://goo.gl/Aej7UF. It is composed of a static body (A, C–E in Fig. 3.1a) that houses three servo motors (B) above the user's fingernail and a mobile platform (G) that applies the requested stimuli to the fingertip. Three cables (H) connect the two platforms, and springs (F) around the cables keep the mobile platform in a reference configuration, away from the fingertip, when not actuated. By controlling the cables length, the motors can orient and translate the mobile platform in three-dimensional space. The device fastens to the finger with a fabric strap fixed to part D.

The actuators used in our prototype are HS-55 MicroLite (Hitec RCD Korea, KR) servo motors. They are able to exert up to 120 N· mm torque. The whole mechanical support is realized using a special type of acrylonitrile butadiene styrene, called ABS*Plus* (Stratasys Inc., USA). The total weight of the device, including actuators, springs, wires, and mechanical support, is 45 g. Note that the force applied by the device to the user's finger pad is balanced by a reaction force supported by the body of the device (E). This structure has a larger contact surface with respect to the mobile platform (G), so that the local pressure is lower and the contact is thus mainly perceived on the finger pad and not on the nail side of the finger [4–6]. Further information about the device's technical specifications are shown in Table 3.1. This 3-DoF cutaneous device has been preliminary presented in [7].

3.2.1 Related Works on Fingertip Compliance Models

This fingertip cutaneous device has been designed to simulate contact interaction forces with virtual objects and surfaces. In order to understand how to define such stimuli in terms of actuation parameters (i.e., input commands for the motors), it is necessary to analyze the fingertip from a mechanical point of view and discuss the relationship between forces and skin deformation. In this section we briefly summarize the main literature on the topic.

Srinivasan and Dankekar [8] presented a 2D continuum fingertip model in which the finger is approximated by an homogeneous, isotropic and incompressible elastic material. Serina et al. [9] developed a model that took into account both inhomogeneity and geometry of the fingertip. The undeformed fingertip was modelled as an axial symmetric ellipsoidal elastic membrane, filled with a incompressible fluid with an internal pressure. This model resulted in a pulp force/displacement relationship that can be represented as a non linear hardening spring, i.e. whose stiffness increases with the applied load. Most of the displacement is reached when the load is 1 N, which corresponds to a displacement of about 2 mm. Cook et al. [10] presented an experimental method to obtain two-dimensional skin tension/extension-ratio characteristics of living human skin. This method incorporates a strain-gauged pretension device, used to determine the naturally existing tension and deformation fields in a specific skin area, and a suction-cup system, which applies a homogeneous plane strain field to the skin at controlled straining rates. Srinivasan et al. [11] experimentally determined the in vivo compressibility of the human fingertip, developing a technique to cast 3D copies of the undeformed fingertip and measure its initial volume. Serina et al. [12] characterized the response of the in vivo fingertip pulp under repeated and compressive loadings, aiming at a better understanding of the pulp force modulation. Their results suggest that high-frequency forces of small magnitude are attenuated by the non-linearly stiffening pulp, while high-frequency forces of larger magnitude are transmitted to the bone. Pulp response was also significantly influenced by the loading angle. Nakazawa et al. [13] studied the force/deformation behavior of the fingertips in the lateral direction. They experimentally measured the impedance characteristics of the fingertips in the direction tangential to the tip surface. The simplified Kelvin model was adopted to describe the relationship between applied tangential force and finger deformation. The impedance characteristics of human fingertips were experimentally identified. Results showed different stiffness properties in the shearing direction: the thumb, for example, was found to be stiffer than the other fingers. Furthermore, the shearing stiffness depended on the direction of the force: fingers were found stiffer in the pointing direction than in the lateral one. Wang and Hayward [14] found out that the stress/strain behavior of the fingertip under shearing forces cannot be considered linear. They experimentally quantified the anisotropic and hysteretic behavior of fingertip deformation under the application of tangential forces. More recently, Wiertlewski and Hayward [15] measured the mechanical impedance of the index fingertip under different loading conditions. They also measured the corresponding Young modulus and viscosity for the finger pad.

3.2.2 Device Model and Control

This section summarizes the device working principles and analyzes the range of forces the device can apply to the fingertip. For the applicable forces, two types of constraints are taken into account: the first one is due to the friction between the finger pad and the platform, while the second one is due to the device mechanical structure and actuation system. To define the subspace of forces the device can generate, we assume that the system, composed of the device and the fingertip, is in static equilibrium conditions. We then neglect platform displacement due to force variations. In fact, in order to take into account for the effect of this displacement, we would need to know more about the fingertip compliance features, which are not easy to evaluate and are subject-related [15].

Device geometry and forces. Similarly to the wearable device presented by Prattichizzo et al. [4], our cutaneous system can be modelled as a 3-DoF parallel mechanism, where the static body is fixed to the nail side of the finger and the mobile platform is in contact with the finger pulp. The platform is actuated by controlling the length and tension of three cables connected to its vertices. When no torques are applied to the motors, three springs maintain the platform in a reference configuration, constraining its motion in the normal and tangential directions. However, in the following quasi-static analysis, for the sake of simplicity, we neglect the springs stiffness in the lateral direction. The mechanical model of this device differs from the one described by Prattichizzo et al. [4], since, in our case, cable directions are not influenced by the finger geometry and curvature. Thanks to a larger mobile platform, cables never touch the finger and, therefore, no problems due to the friction between cables and skin arise.

With reference to Fig. 3.2a, b, B_1, B_2, and B_3 indicate the points, on the mobile platform, where the cables, linking the mobile platform to the three motors, pass. $S_1 = \langle O_1, x, y, z \rangle$ is a reference frame on the mobile platform, whose origin O_1 is placed at the platform's geometric center and whose z axis is orthogonal to the plane defined by B_i, pointing towards the fingertip. A_1, A_2, and A_3 indicate the points, on the static body of the device, where the cables pass, and $S_0 = \langle O_0, x_0, y_0, z_0 \rangle$ is a

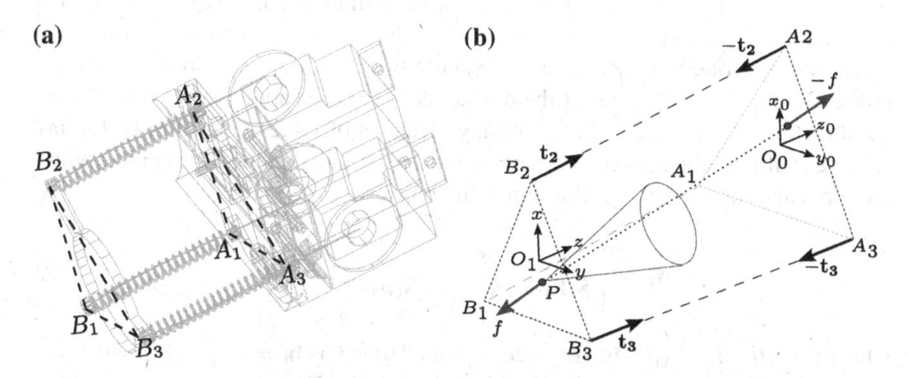

Fig. 3.2 The 3-DoF cutaneous device model. In **a**, A_i and B_i are the points where the cables link the static and mobile parts of the device, respectively. In **b**, force **f** is applied by the fingertip to the device at point P. Its direction has to be inside the friction cone (in *red*) in order to fulfil our friction constraint. On the other hand, $\mathbf{t_i}$ represent the forces applied by the actuators to the mobile platform. **a** Device kinematic scheme. **b** Forces acting on the device (color figure online)

reference frame on that body, whose origin is located at O_0, and whose z_0 axis is orthogonal to the plane defined by A_i.

Wrench $\mathbf{w_p} = [\mathbf{f}^T, \mathbf{m}^T]^T$, applied by the fingertip on the platform and expressed w.r.t. point P, is balanced by the forces $\mathbf{t_1}$, $\mathbf{t_2}$, and $\mathbf{t_3}$ applied by the three cables (see Fig. 3.2b). Point P can be defined as the intersection between the central axis of the stress distribution and the plane defined by points B_i. Even though the contact between the fingertip and the mobile platform is approximately at the center of the mobile platform, for the sake of generality, we consider $P \neq O_1$.

Contact and friction. The ratio between the tangential and the normal component of the contact force $\mathbf{f}$, in classical friction models, is limited by the friction coefficient μ, which depends on the surface material and conditions. This constraint has a meaningful graphical interpretation, that is shown in Fig. 3.2b: the direction of the contact force $\mathbf{f}$ has to lie within a cone, (i) whose axis is the normal direction to the contact surface at the (theoretical) contact point, (ii) whose vertex is point P itself, and (iii) has a vertex angle of $\varphi = \arctan \mu$. This cone is usually referred to as a *friction cone* [16].

Device equilibrium. The directions of forces $\mathbf{t_1}, \mathbf{t_2}$, and $\mathbf{t_3}$ are defined, respectively, by the unitary vectors $\mathbf{s_1}$, $\mathbf{s_2}$, and $\mathbf{s_3}$, which go from the mobile platform to the static body. These forces can be thus expressed as the sum of two components: the first one represents the force applied by the ith actuator and its module depends on the motor torque, while the second one represents the contribution generated by the spring and its module depends on the spring stiffness and deformation. Forces $\mathbf{t_i}$, with $i = 1, \ldots, 3$ can be then expressed as

$$\mathbf{t_i} = t_i \mathbf{s}_i, \quad t_i = \left(\frac{\eta_i}{r_i} - k_i \left(|d_i| - |d_{i,o}| \right) \right), \tag{3.1}$$

where η_i is the torque of the ith motor, r_i is the ith motor pulley radius ($r_i = 8$ mm in our case), k_i is the spring stiffness ($k_i = 0.7$ N/mm in our case), $|d_i|$ the actual cable length, and $|d_{i,0}|$ the nominal spring length ($|d_{i,0}| = 20$ mm in our case).

Beside the friction cone, which substantially depends on the properties of the contact surface, the structure of the device itself imposes additional constraints on the direction of the contact force. For any given set of forces $\mathbf{t_1}$, $\mathbf{t_2}$, and $\mathbf{t_3}$, for any relative configuration between the parts of the device, and for any given contact point, the force and momentum equilibrium equations lead to

$$\mathbf{w_p} = \begin{bmatrix} \mathbf{s_1} & \mathbf{s_2} & \mathbf{s_3} \\ \mathbf{S(b_1)} & \mathbf{S(b_2)} & \mathbf{S(b_3)} \end{bmatrix} \mathbf{t} = \mathbf{A}\,\mathbf{t}, \tag{3.2}$$

where $\mathbf{t} = [t_1, \ t_2, \ t_3]^T$, $\mathbf{b_i}$ are three vectors defined as $\mathbf{b_i} = (B_i - P)$, and $S(\mathbf{b_i})$ denotes the skew matrix associated with the vectors $\mathbf{b_i}$. For any given configuration, the possible wrenches that can be generated are therefore those belonging to the column space of matrix $\mathbf{A}$. However, it is not possible to define an inverse relationship between wrenches and forces. In other terms, since the device is underactuated, it is not possible to arbitrarily set the wrench $\mathbf{w_p}$ and, from it, evaluate $\mathbf{t_i}$.

Maximum force inclination. Assuming $t_i \geq 0$ and considering the structure of the device for any contact point and any device configuration, it is easy to assess, by considering Eq. (3.2), that the maximum inclination ϕ_{max} of force $\mathbf{f}$, w.r.t. the normal direction identified by the z axis, is given by

$$\phi_{max} = \max_{i=1,2,3} (\arcsin |\mathbf{s_i} \times \mathbf{k}|), \tag{3.3}$$

where $\mathbf{k}$ denotes the unit vector parallel to the z axis. This is another constraint for the contact force the device can apply to the fingertip, and it only depends on the geometry and structure of the device. Also, this second constraint has a graphical interpretation: the force $\mathbf{f}$ has to stay inside a pyramidal surface with a triangular basis, whose vertex is in P and whose lateral edges are defined by the directions parallel to $\mathbf{s_1}$, $\mathbf{s_2}$, and $\mathbf{s_3}$. The range of applicable force directions thus depends on the inclination of the $\mathbf{s_i}$ vectors with respect to the platform. In order to be able to apply a wider range of forces, i.e., higher tangential forces with respect to the normal ones, it is necessary to increase these inclinations. For example, we could increase the size of the static body, the one placed on the nail side of the finger, and/or shape the device to reduce the distance between points A_i and B_i in the z direction.

Design for stability. Unlike other interfaces designed to apply solely shear stresses [17, 18], our cutaneous device can provide only forces with a maximum inclination ϕ_{max}. This means that a component of the force normal to the fingertip surface will always be present and clearly perceivable. Although this limits the actuation capabilities of our device, it also suggests a very interesting mechanical property. A force applied by the mobile platform to the fingertip will be always compensated by a reaction force supported by the static body of the device (called E in Fig. 3.1a). This reaction force is balanced by the contact with the finger nail and by the joints of the grounded haptic device (see Fig. 3.1b). The ratio between these two force components depends on the fingertip and Omega compliances, both evaluated with respect to the point in which the cutaneous device is located. We assume that the part of the reaction force applied to the Omega is much smaller with respect to the other forces involved. In the hypothesis that this force component can be neglected, this type of cutaneous feedback does not affect the stability of the teleoperation system in which it is employed, since the force is provided *only* at the fingertip level and does not influence the position of the master end-effector. The validity of this assumption is confirmed by the experiments described in Sects. 3.3.1 and 3.3.3.

Contact stiffness. It is also important to consider that the platform, by applying a wrench $-\mathbf{w_p}$, deforms the fingertip. Any displacement of the platform $\boldsymbol{\xi} = [p_x \; p_y \; p_z \; \alpha \; \beta \; \gamma]^T$ when in contact with the finger, leads to a contact stress distribution on the finger pad. The resultant force and moment of the normal and tangential stress distributions, arising at the contact patch, balance the external wrench $-\mathbf{w_p}$. Fingertip deformation and applied wrench can be related by an impedance model, that is necessarily non-linear and depends on the fingertip specific characteristics (e.g., geometric parameters, subject's age). The use of a fingertip impedance model is necessary to solve the kinematics indeterminacy caused by the underactuation of our cutaneous system. In this work we assume a simplified linear relationship between the resultant wrench and the platform displacement. In other terms, we consider the platform configuration $\boldsymbol{\xi}$ proportional to the wrench $\mathbf{w_p}$,

$$\boldsymbol{\xi} = \mathbf{K}^{-1}\mathbf{w_p}, \tag{3.4}$$

where $\mathbf{K} \in \mathbb{R}^{6\times6}$ is the fingertip stiffness matrix, as defined in [19].

Control. From the control point of view, the device can be represented as a non-linear, multi-input multi-output (MIMO) coupled system. Furthermore, the system is underactuated, since three motors control the motion of the platform, which has six degrees of freedom [4]. Several control strategies can be thus considered. For example, we can control the force applied by the platform to the fingertip, or the position and orientation of the mobile platform. The applied forces and the platform configuration are related by the aforementioned compliance model and by the stiffness of the springs. It is important to highlight that before starting to use the device, regardless of the control strategy adopted, the initial position of the mobile platform must be calibrated according to the size of the user's fingertip.

3.3 Experimental Evaluation

In order to evaluate the feasibility of the sensory subtraction approach in a complex teleoperation scenario, we carried out two experiments, both in virtual and real environments. The first experiment consists of a peg-in-hole task in a virtual environment.

3.3.1 Experiment #1: Teleoperation in a Virtual Environment

3.3.1.1 Experimental Setup and Methods

Figure 3.3 shows the experimental setup. The master system is composed of two cutaneous devices, attached to the end-effectors of two Omega 3 haptic interfaces, as also shown in Fig. 3.1b. The virtual environment is composed of a peg, a board with two holes (named $hole_1$ and $hole_2$ in Fig. 3.3b), and two small spheres.

The holes are 35 mm deep (x-direction), 35 mm wide (y-direction), and 5 mm high (z-direction). The peg is a 150 g cube with an edge length of 30 mm, and, therefore, the hole has a clearance of 5 mm in the x and y directions. The spheres have a diameter of 10 mm, and their position is linked to the virtual location of the subject's fingers, evaluated according to the popular god-object model [20] (i.e., the spheres are placed where the haptic interface end-effector would be if the haptic interface and the objects were infinitely stiff). A spring $k = 1$ N/mm is used to model the contact force between the spheres, controlled by the user, and the virtual objects. Subjects

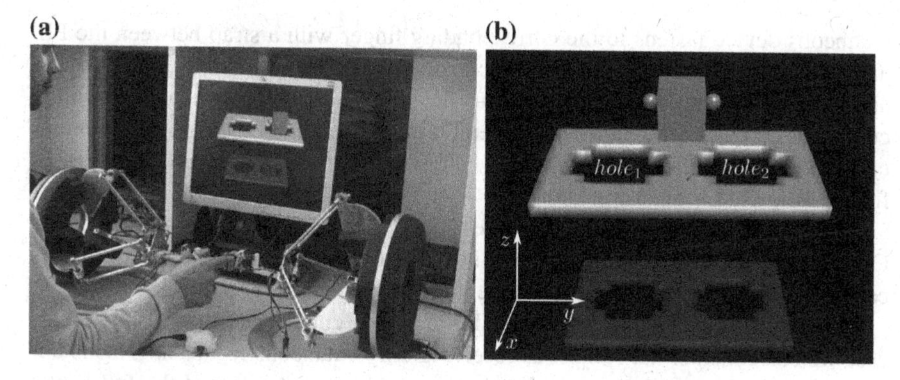

Fig. 3.3 Teleoperation in a virtual environment. Subjects are required to wear two cutaneous devices, one on the thumb and one on the index finger, and then complete the peg-in-hole task as fast as possible. **a** Master system. **b** Virtual environment

are able to interact with the virtual environment using CHAI 3D, an open-source set of C++ libraries for computer haptics and interactive real-time simulation.

3.3.2 Methods

The task consisted of grasping the cube from the ground, inserting it into the right hole ($hole_2$), then in the left hole ($hole_1$), and then again in $hole_2$ and $hole_1$. The task started when the subject touched the object for the very first time and finished when the subject inserted, for the second time, the peg in $hole_1$. At least half of the length of the peg had to be inserted in each hole, from the top to the bottom. When the object was correctly inserted into the hole, the color of the peg changed. A video of the experiment is available as supplemental material at http://extras.springer.com/978-3-319-25455-5 and at http://goo.gl/ybV59w. In order to make the task easier to complete, the angular velocity of the peg was set to zero (i.e., the peg was not allowed to rotate).

Each participant made sixteen trials of the peg-in-hole task, with four randomized repetitions for each feedback condition proposed:

- kinesthetic and cutaneous feedback, provided by both the Omega and the cutaneous devices (full haptic feedback, condition $(K+C)_v$),
- kinesthetic feedback only, provided by the Omega interfaces (condition K_v),
- cutaneous feedback only, provided by the cutaneous devices (sensory subtraction approach, condition C_v),
- no haptic feedback (condition N_v).

In condition $(K+C)_v$, the force computed by the virtual environment is fed back by both the Omega interfaces and the cutaneous devices, assuming the two haptic stimuli to be provided independently from each other. In this case, in fact, each cutaneous device fastens to the corresponding finger with a strap between the PIP and DIP joints. For this reason, we can consider the force provided by the Omega as applied only at that point. The cutaneous device is thus the only display providing cutaneous stimuli to the fingertip. Nonetheless, of course, the Omega still provides cutaneous stimuli at the point where the strap is fastened, but we can consider this force negligible for our purposes.

In condition K_{υ}, the cutaneous devices are not active: the force is provided by the Omega only, and the finger pulp is not in contact with the mobile platform. In this case we can consider the subject as receiving only kinesthetic feedback.

In condition C_{υ}, the force is provided through the cutaneous devices only. In this case we can consider the subject as receiving only cutaneous feedback. This is our sensory subtraction technique, as it subtracts the kinesthetic part of the full haptic interaction to leave only cutaneous cues.

In condition N_{υ}, no haptic feedback is provided to the subjects.

In all the conditions, the Omega interfaces are in charge of tracking the position of the fingers. Visual feedback, as shown in Fig. 3.3, is always provided to the subjects.

In order to determine the number of subjects needed for our research study, we ran a power analysis using the G*Power software. We estimated the effect size from the data retrieved in [7]. Power analysis revealed that, in order to have a 95 % chance of detecting differences among the metrics taken into account (completion time and exerted forces), we would need at least 15 participants (completion time: effect size 1.30, actual power 0.952; exerted forces: effect size 1.31, actual power 0.957).

Fifteen participants took part in the experiment, including 10 males and 5 females. Seven of them had previous experience with haptic interfaces, but only three had tried cutaneous devices before. None of the participants reported any deficiencies in their visual or haptic perception abilities, and all of them were right-hand dominant. Participants were briefed about the task and afterward signed an informed consent, including the declaration of having no conflict of interest. All of them were able to give the consent autonomously. The participation in the experiment did not involve the processing of genetic information or personal data (e.g., health, sexual lifestyle, ethnicity, political opinion, religious or philosophical conviction), neither the tracking of the location or observation of the participants.

Participants were provided with a 10-min familiarization period with the experimental setup. They were then asked to wear two cutaneous devices, one on the thumb and one on the index finger (see Fig. 3.3a), and complete the peg-in-hole task as fast as possible.

3.3.2.1 Results

In order to evaluate the performance of the considered feedback conditions, we recorded (1) the time needed to complete the task and (2) the forces generated by the contact between the two spheres, controlled by the subject, and the cube. Data resulting from different repetitions of the same condition, performed by the same subject, were averaged before comparison with other conditions' data. All the subjects were

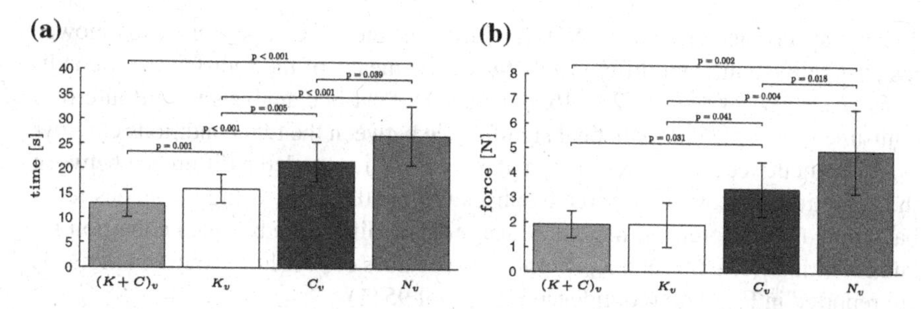

Fig. 3.4 Teleoperation in a virtual environment. Completion time of the peg-in-hole task and force generated by the contact between the spheres and the object are plotted (mean and standard deviation) during tests with both kinesthetic and cutaneous force feedback (task $(K + C)_v$), kinesthetic only (task K_v), cutaneous only (task C_v), and no force feedback (task N_v). P-values of post-hoc group comparisons are reported when a statistical difference is present (confidence interval of 95 %). **a** Completion time. **b** Contact force

able to complete the task. In the following analysis, completion time and exerted forces are treated as dependent variables, while the feedback condition is treated as the independent variable.

Figure 3.4a shows the average time elapsed between the instant the subject grasped the cube for the very first time and the instant he or she completed the peg-in-hole task. The collected data passed the Shapiro–Wilk normality test (see Table 3.2 for details) and the Mauchly's Test of Sphericity ($\chi^2(5) = 10.427$, $p = 0.065$). Comparison of the means among the feedback condition was tested using a one-way repeated measures ANOVA ($F(3, 42) = 48.410$, $p < 0.001$). The means differed significantly. Post-hoc analysis with Bonferroni adjustments revealed statistically significant difference between all the groups, showing that the time needed to complete the task depends on the feedback condition. Significant p-values of post-hoc group comparisons are reported in Fig. 3.4a (confidence interval of 95 %).

Figure 3.4b shows the average contact forces generated between the two spheres, controlled by the subject, and the cube along the y-direction, i.e., the one perpendicular to the cube surface (see Fig. 3.3b). The collected data passed the Shapiro–Wilk normality test (see again Table 3.2). Mauchly's Test of Sphericity indicated that the assumption of sphericity had been violated ($\chi^2(5) = 16.988$, $p = 0.005$).

Table 3.2 Shapiro–Wilk normality test (virtual scenario)

Feedback condition	Completion time			Contact force		
	Statistic	df	Sig.	Statistic	df	Sig.
$(K + C)_v$	0.911	15	0.138	0.921	15	0.198
K_v	0.949	15	0.510	0.938	15	0.353
C_v	0.953	15	0.572	0.925	15	0.233
N_v	0.957	15	0.632	0.917	15	0.174

A one-way repeated measures ANOVA with a Greenhouse-Geisser correction showed a statistically significant difference between the means of the four feedback conditions ($F(1.704, 23.861) = 27.046$, $p < 0.001$). Post-hoc analysis with Bonferroni adjustments revealed no statistical significance between the two conditions employing kinesthetic feedback ($(K+C)_v$ and K_v), while it revealed a difference between the condition employing no force feedback (N_v) and the one using cutaneous feedback only (C_v). Moreover, there was a significant difference between condition C_v and conditions $(K+C)_v$ and K_v. Significant p-values of post-hoc group comparisons are reported in Fig. 3.4b (confidence interval of 95 %).

Figure 3.5 reports the position of the cube along the z- and y-axes. Average trajectory of the peg along the z- and y-axes (solid green line) ± standard deviation (green patch) along the z-axis is shown for each feedback condition. The size of the green patch gives a measure of the variability of the trajectory among the subjects.

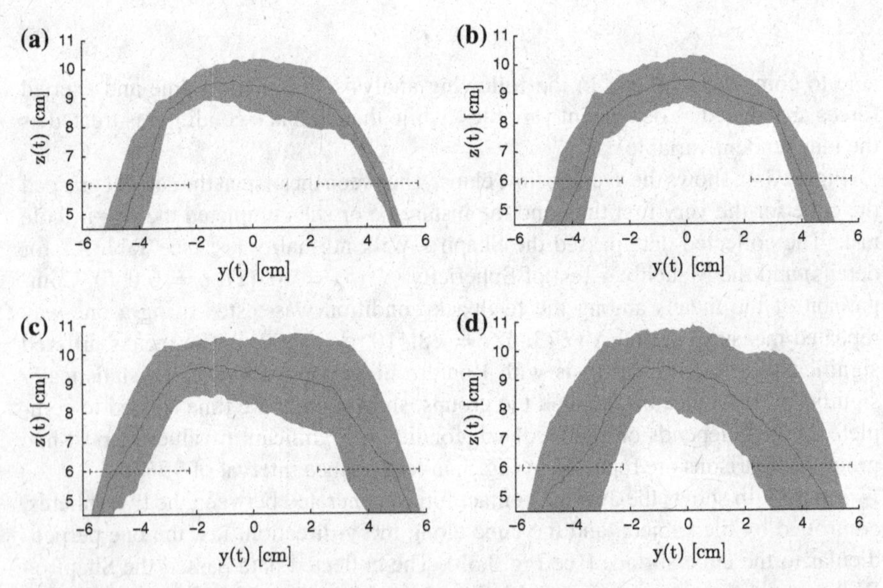

Fig. 3.5 Teleoperation in a virtual environment. Average trajectory of the peg along the z- and y-axes (*solid green line*) ± standard deviation (*green patch*) along the z-axis is shown for each feedback condition. The position of the two holes (*dashed red lines*) are reported as well. Subjects were able to complete the peg-in-hole task in all the considered conditions. The size of the *green patch* gives a measure of the variability of the trajectory among the subjects. **a** Kinesthetic and cutaneous feedback, $(K+C)_v$. **b** Kinesthetic force feedback, K_v. **c** Cutaneous force feedback, C_v. **d** No force feedback, N_v (color figure online)

3.3.3 Experiment #2: Teleoperation in a Real Environment

The second experiment consists in a peg-in-hole experiment in a real environment. For the sake of conciseness, this time we focused our attention only on feedback conditions providing cutaneous cues.

3.3.3.1 Experimental Setup and Methods

Figures 3.6 and 3.7 show the experimental setup. The master system again consists of two cutaneous devices attached to the end-effectors of two Omega 3 haptic interfaces. The slave system is composed of a 6-DoF manipulator KUKA KR3 and a DLR-HIT Hand II. The robotic hand is attached to the end-effector of the manipulator. A video camera is placed near the manipulator to let the operator see the hand and its nearby environment. The environment is composed of a cylinder of 150 g and a rigid board with two circular holes (see Fig. 3.6b). The holes have a diameter of 45 mm and are 35 mm high. The cylinder has a base radius of 35 mm and a height of 200 mm. Therefore, the hole has a radial clearance of 10 mm in the x-y plane. In this second experiment we used a cylinder so that peg's rotations about the z-axis did not prevent it from entering the holes, making the task easier to complete.

In order to ensure a convincing illusion of telepresence, the movements of the operator should be well replicated by the slave system [21]. In our system the average position of the operator's fingers controls the telemanipulator wrist, and the distance between the two fingers is linked to the distance between the thumb and index fingers of the robotic hand.

With reference to Fig. 3.7a, let $\{\mathcal{L}\}$ and $\{\mathcal{R}\}$ be two reference frames, whose origins are located at the geometric center of the base of the left and right haptic interfaces, respectively. Points $\mathbf{p}_\mathcal{L}, \mathbf{p}_\mathcal{R} \in \mathbb{R}^3$ represent the positions of the haptic devices'

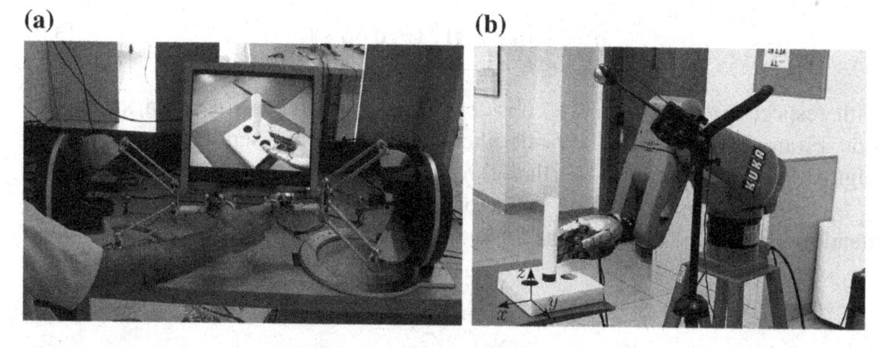

Fig. 3.6 Teleoperation in a real environment. The master side is composed of two cutaneous devices fixed to the end-effectors of two Omega 3 haptic devices, while the slave side consists of the DLR-HIT Hand II attached to the 6 DoFs manipulator KUKA KR3. **a** Master system. **b** Slave system

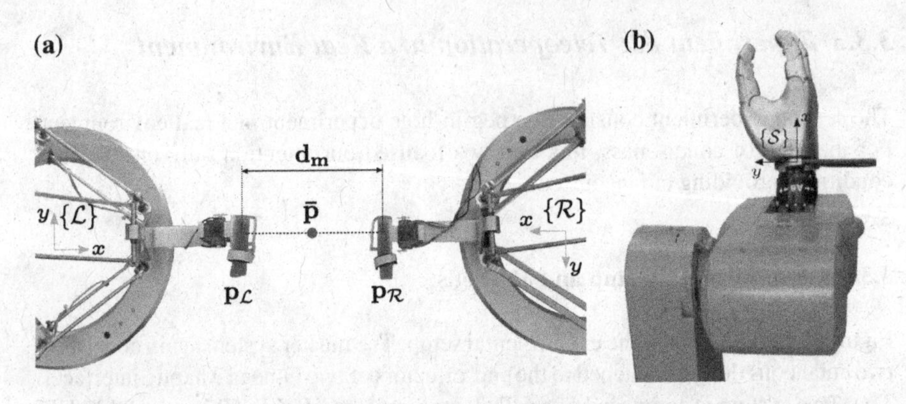

Fig. 3.7 Teleoperation in a real environment. $\{\mathcal{L}\}$ and $\{\mathcal{R}\}$ are two reference frames on the Omega haptic interfaces and $\bar{\mathbf{p}}$ is the point located between the two end effectors. **b** shows the origin of the slave reference frame $\{\mathcal{S}\}$, which is placed on the robotic hand wrist. **a** Master system. **b** Slave system

end-effectors with respect to $\{\mathcal{L}\}$ and $\{\mathcal{R}\}$, respectively. Let $\mathbf{d} = [d_x \; d_y \; d_z]^T$ be a vector, expressed w.r.t. the $\{\mathcal{L}\}$ frame, representing the distance between the origins of $\{\mathcal{L}\}$ and $\{\mathcal{R}\}$, $\mathbf{R_z}(\pi) \in \mathbb{R}^{3\times3}$ the π rotation matrix about the z-axis, and

$$\mathbf{H}_{\mathcal{L}}^{\mathcal{R}}(\pi, \mathbf{d}) = \begin{bmatrix} \mathbf{R_z}(\pi) & \mathbf{d} \\ \mathbf{0} & 1 \end{bmatrix} \tag{3.5}$$

the homogeneous transformation matrix from $\{\mathcal{R}\}$ to $\{\mathcal{L}\}$. Note that, due to the relative positioning of the haptic interfaces and reference frames, only the first component of the distance vector, d_x, is different from zero. The correct positioning of the haptic devices was carefully checked before the beginning of each experiment. The average position of the fingers can be then expressed as

$$\tilde{\bar{\mathbf{p}}} = \frac{1}{2}\left(\tilde{\mathbf{p}}_{\mathcal{L}} + \mathbf{H}_{\mathcal{L}}^{\mathcal{R}}(\pi, \mathbf{d})\, \tilde{\mathbf{p}}_{\mathcal{R}}\right), \tag{3.6}$$

with respect to $\{\mathcal{L}\}$, which we consider our base reference frame on the master side. From now on, we will consider the notation $\tilde{\mathbf{k}}$ to represent the vector $\mathbf{k} \in \mathbb{R}^3$ augmented by appending a "1", therefore $\tilde{\mathbf{k}} = [\mathbf{k}^T \quad 1]^T$.

Since the telemanipulator end-effector was controlled in velocity, the input command $\mathbf{v} \in \mathbb{R}^3$ can be expressed as

$$\mathbf{v} = \mathbf{R_z}\left(\frac{\pi}{2}\right) \dot{\bar{\mathbf{p}}}, \tag{3.7}$$

where $\dot{\bar{\mathbf{p}}} \in \mathbb{R}^3$ is the velocity of the midpoint $\bar{\mathbf{p}}$, located between the end effectors of the two Omega haptic interfaces, and $\mathbf{R_z}(\frac{\pi}{2}) \in \mathbb{R}^{3\times3}$ is the $\frac{\pi}{2}$ counter-clockwise rotation matrix about the z-axis, which performs the rotation from the master reference system $\{\mathcal{L}\}$ to the slave reference frame $\{\mathcal{S}\}$ (see Fig. 3.7b). For the sake of simplicity, and due to the nature of the telemanipulation task, we considered the orientation of the robotic wrist fixed.

For the robotic hand, the distance between the subject's fingers controlled the distance between the robotic ones. Let

$$\tilde{\mathbf{d}}_m = \tilde{\mathbf{p}}_{\mathcal{L}} - \mathbf{H}_{\mathcal{L}}^{\mathcal{R}}(\pi, \mathbf{d})\, \tilde{\mathbf{p}}_{\mathcal{R}} \tag{3.8}$$

be the distance between the two master's end-effectors, expressed w.r.t. $\{\mathcal{L}\}$. Let us also define n_q as the number of joints of the robotic fingers and

$$\mathbf{d_s} = \mathbf{R_z}(\pi)\, \mathbf{d_m} \tag{3.9}$$

as the distance between the master's end-effectors w.r.t. $\{\mathcal{S}\}$. For the sake of simplicity, and due to the nature of the telemanipulation task, we took into account only movements along the y direction (with respect to $\{\mathcal{S}\}$). Changes in $\mathbf{d_s}$ controlled the velocity of the robotic thumb as

$$\dot{\mathbf{d}}_t = \frac{\dot{\mathbf{d}}_s}{2}, \tag{3.10}$$

and of the robotic index finger as

$$\dot{\mathbf{d}}_i = \begin{pmatrix} 1 & 0 & 0 \\ 0 & -1 & 0 \\ 0 & 0 & 1 \end{pmatrix} \frac{\dot{\mathbf{d}}_s}{2}. \tag{3.11}$$

The velocities of the robotic hand joints $\dot{q} \in \mathbb{R}^{2n_q}$ can thus be expressed as

$$\dot{\mathbf{q}} = \mathbf{J}^{\#}\dot{\mathbf{d}}_r\, \zeta, \tag{3.12}$$

where $\mathbf{J}^{\#}$ is the Moore–Penrose pseudoinverse of the robotic hand Jacobian matrix $\mathbf{J} \in \mathbb{R}^{6\times 2n_q}$, $\dot{\mathbf{d}}_r = [\dot{\mathbf{d}}_t\ \ \dot{\mathbf{d}}_i]^{\mathrm{T}}$, and ζ is the scaling factor between the master and slave workspace. Due to the mechanical design of each robotic finger, which has only three joints, we assumed the robotic hand as not redundant, i.e., the null space $\mathcal{N}(\mathbf{J})$ is empty.

Similarly, contact forces $\boldsymbol{\lambda} \in \mathbb{R}^6$, exerted by the hand at the two contact points, can be expressed as

$$\boldsymbol{\lambda} = (\mathbf{J}^{\mathrm{T}})^{\#}\boldsymbol{\tau} + \mathbf{N}_{\mathbf{J}^{\mathrm{T}}}\, \boldsymbol{\chi}, \tag{3.13}$$

where $\mathbf{N}_{\mathbf{J}^{\mathrm{T}}}$ is a matrix whose columns form a basis of $\mathcal{N}(\mathbf{J}^{\mathrm{T}})$, and the vector $\boldsymbol{\chi}$ parametrizes the homogeneous solution to the equilibrium problem $\boldsymbol{\tau} = \mathbf{J}^{\mathrm{T}}\boldsymbol{\lambda}$. In the literature, generic contact forces that satisfy the condition $\mathbf{J}^{\mathrm{T}}\mathbf{N}_{\mathbf{J}^{\mathrm{T}}}\, \boldsymbol{\chi} = \mathbf{0}$ are referred as structural forces [22]. Since we are interested in feeding back only forces due to the grasp, we filtered out the ones registered during free-hand movements.

The system is managed by a GNU/Linux machine, equipped with a real-time scheduler. It communicates via UDP/IP with the controller of the robotic hand and via Eth.RSIXML (KUKA Roboter GmbH, Germany) with the telemanipulator. The haptic and cutaneous devices use their own embedded microcontrollers and are connected to the GNU/Linux machine via USB. The master and slave systems are connected to the same Local Area Network (LAN). However, thanks to the underlying communication infrastructure, they could have been also easily placed in different LANs and then communicate through an Internet connection.

So as to preserve the stability of the teleoperation system, we took into account a popular passivity controller, able to regulate the force fed back to the operator. In particular, we employed the two-layer technique described by Franken et al. [23]. The control architecture is split into two separate layers. The hierarchical top layer, named *Transparency Layer*, aims at achieving the desired transparency, while the lower layer, named *Passivity Layer*, ensures the passivity of the system. Separate communication channels connect the layers at the slave and master levels so that information related to exchanged energy is separated from information about the desired behavior. Note that the objective of this work is not the design of an efficient passivity controller, but the validation of the proposed cutaneous approach. We used a stability controller *only* to be able to compare full haptic feedback with our cutaneous-only sensory subtraction technique.

The task consisted of grasping the cylinder from its initial position (see Fig. 3.6b) and inserting it into two holes, following a sequence of insertion analogous to the one already described in Sect. 3.3.2.1. The task started when the subject touched the object for the first time and finished when the subject inserted, for the second time, the peg in the hole closer to the manipulator. At least half of the length of the peg had to be inserted in the hole, from the top to the bottom. The grasp had to be performed using the fingertip of the robotic hand. A video of the experiment is available as supplemental material at http://extras.springer.com/978-3-319-25455-5 and at http://goo.gl/12juj8.

Each participant made twelve trials of the peg-in-hole task, with four randomized repetitions for each force feedback condition proposed:

- kinesthetic and cutaneous feedback, provided by the Omega and the cutaneous devices, employing the passivity controller described at the end of Sect. 3.3.3.1 (full haptic feedback, condition $(K + C + P)_r$),
- kinesthetic and cutaneous feedback, provided by the Omega and the cutaneous devices (full haptic feedback, no stability controller, condition $(K + C)_r$),
- cutaneous feedback only, provided by the cutaneous devices (sensory subtraction approach, condition C_r).

In condition $(K+C+P)_r$, the contact force $\boldsymbol{\lambda}$ registered at the robotic fingers (see Eq. 3.13), is fed back by both the Omega interfaces and the cutaneous devices. The passivity controller guarantees the stability of the system. This represents our ideal condition, when the subject is provided with full haptic feedback and no unstable behavior arise.

Condition $(K+C)_r$ is similar to condition $(K+C+P)_r$. The contact force $\boldsymbol{\lambda}$ is again fed back by both the Omega interfaces and the cutaneous devices. However, this time, no passivity algorithm guarantees the stability of the teleoperation loop.

In condition C_r, the contact force $\boldsymbol{\lambda}$ is provided through the cutaneous devices only. This is our cutaneous-only sensory subtraction condition.

In all the conditions, the Omega interfaces are in charge of tracking the position of the fingers. Visual feedback, as shown in Fig. 3.6a, is always provided to the subjects by a video camera placed next to the manipulator arm.

The same fifteen participants who took part in the experiment of Sect. 3.3.1 participated in this one as well. Participants were briefed about the task and a 10-min familiarization period was provided to make them acquainted with the experimental setup. They were then asked to wear two cutaneous devices, one on the thumb and one on the index finger, and complete the peg-in-hole task as fast as possible.

3.3.3.2 Results

In order to evaluate the performance of the different feedback conditions, we recorded (1) the task success rate, (2) the time needed to complete the task, and (3) the forces generated by the contact between the two spheres, controlled by the subject, and the cylinder. Data resulting from different repetitions of the same condition, performed by the same subject, were averaged before comparison with other conditions' data. In the following analysis, success rate, completion time, and exerted forces are treated as dependent variables, while the feedback condition is treated as the independent variable.

Figure 3.8a shows the average task success rate. A trial was considered *not* successful if the subject was not able to complete the peg-in-hole task (i.e., the peg fell out of the workspace of the slave robot). Success rates for conditions $(K+C+P)_r$, $(K+C)_r$, and C_r were 100 %, (23.3 ± 19.97) %, and 100 %, respectively. The main reason for failing the task during condition $(K+C)_r$ was the presence of large oscillations, due to the lack of any stability control technique (see the video linked above). No time limit was imposed.

For the completion time and the average exerted forces, we analyzed only conditions $(K+C+P)_r$ and C_r, since condition $(K+C)_r$ was clearly unsuitable due to its unstable behavior. Figure 3.8b shows the average time elapsed between the instant the subject touches the object for the first time and the instant he or she completes the peg-in-hole task. The collected data passed Shapiro–Wilk normality test (see Table 3.3 for details). A paired-samples t-test determined that the time needed to complete the task differed statistically significantly between feedback conditions ($(K+C+P)_r$ vs. C_r, $t(14) = -2.381$, $p = 0.032$, confidence interval of 95 %).

Figure 3.8c shows the average forces generated by the contact between the two robotic fingers, controlled by the subject, and the peg along the y-direction of $\{S\}$

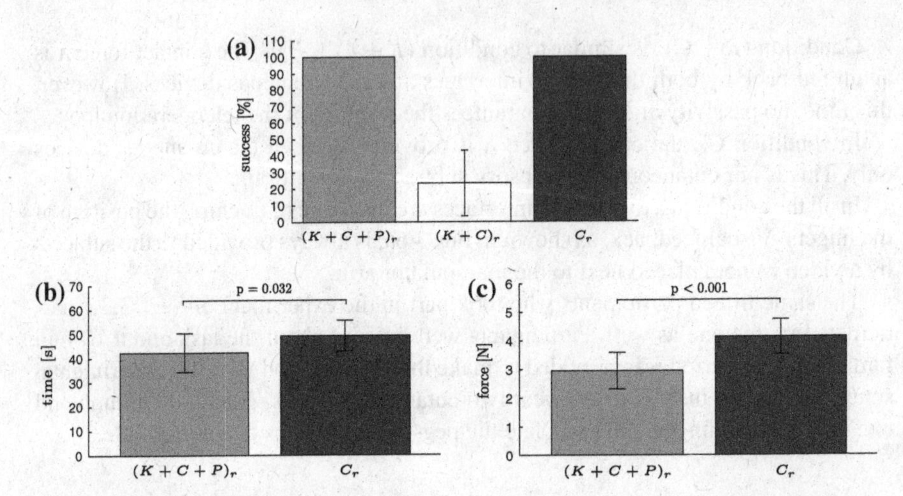

Fig. 3.8 Teleoperation in a real environment. **a** Average task success rate during tests with both kinesthetic and cutaneous feedback (conditions $(K + C + P)_r$ and $(K + C)_r$), and cutaneous feedback only (condition C_r) is plotted (mean and standard deviation). **b** Completion time of the peg-in-hole task and **c** force generated by the contact between the two robotic fingers and the object, with kinesthetic and cutaneous feedback (condition $(K + C + P)_r$), and cutaneous feedback only (condition C_r) are plotted (mean and standard deviation). P-values of paired t-tests are reported in (**b**) and (**c**) when a statistical difference is present (confidence interval of 95 %)

Table 3.3 Shapiro–Wilk normality test (real scenario)

Feedback condition	Completion time			Contact force		
	Statistic	df	Sig.	Statistic	df	Sig.
$(K + C + P)_r$	0.937	15	0.351	0.940	15	0.383
C_r	0.963	15	0.744	0.960	15	0.687

(see Fig. 3.7b). The collected data passed Shapiro–Wilk normality test (see again Table 3.3). A paired-samples t-test determined that the average force exerted differed statistically significantly between feedback conditions ($(K+C+P)_r$ vs. C_r, $t(14) = -5.594$, $p < 0.001$, confidence interval of 95 %).

Figure 3.9 reports the average trajectory of the peg during the task. Trajectories were averaged among subjects for each feedback condition. Average trajectory of the peg along the z- and y-axes (solid blue line) $\pm$ standard deviation (blue patch) along the z-axis is shown for each feedback condition. The size of the blue patch gives a measure of the variability of the trajectory among the subjects.

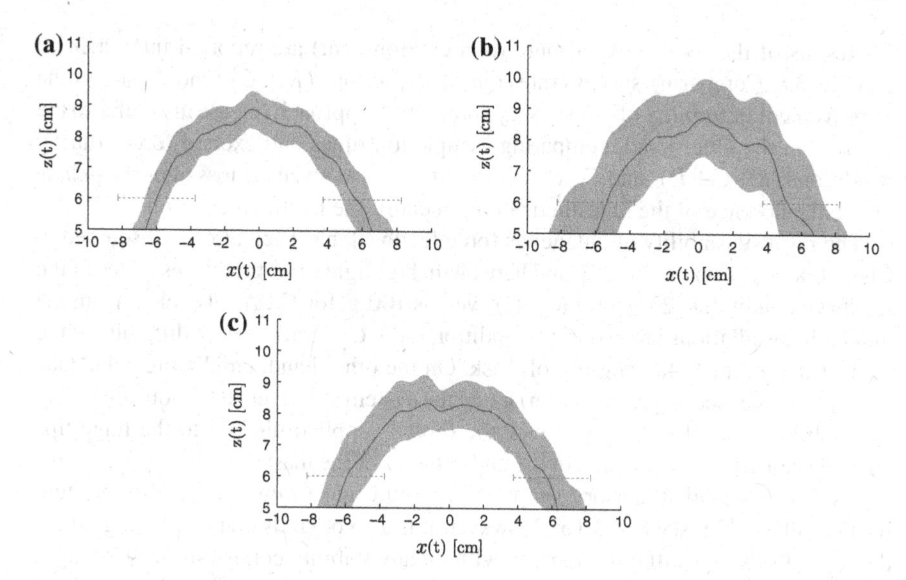

Fig. 3.9 Teleoperation in a real environment. Average trajectory of the peg along the z- and x-axes (*solid blue line*) ± standard deviation (*blue patch*) along the z-axis is shown for each feedback condition. The position of the two holes (*dashed red lines*) are reported as well. The size of the *blue patch* gives a measure of the variability of the trajectory among the subjects. **a** Kinesthetic and cutaneous force feedback, employing the passivity controller, $(K+C+P)_r$. **b** Kinesthetic and cutaneous force feedback, no passivity controller, $(K+C)_r$. **c** Cutaneous force feedback, C_r (color figure online)

3.4 Discussion

We evaluated the sensory subtraction approach in two challenging experimental scenarios, extending the results presented in Chap. 2.

Results of the first experiment (virtual environment) are reported in Sect. 3.3.1 and Fig. 3.4. While receiving kinesthetic and cutaneous feedback (condition $(K+C)_v$), subjects completed the task in less time than while receiving kinesthetic feedback only (condition K_v). Moreover, employing cutaneous feedback only (sensory subtraction, condition C_v) yielded to significant better results than employing no force feedback at all (condition N_v), both in terms of completion time and exerted forces.[1] However, as expected, using kinesthetic feedback (both conditions $(K+C)_v$ and K_v) still showed better performance with respect to employing either cutaneous feedback only or no force feedback at all. Nonetheless, we believe that this reduction of performance is a price worth paying in order to get a great improvement in the stability of the teleoperation loop, as it is clear from the second experiment.

[1] Note that a higher force fed back to the subjects meant a larger penetration into the virtual object and a higher energy expenditure during the grasp. Measuring the average of intensities of the contact forces is a widely-used approach to evaluate energy expenditure during the grasp [22].

Results of the second experiment (real environment) are reported in Sect. 3.3.3 and Fig. 3.8. Comparing success rates during conditions $(K+C)_r$ and C_r shows the improvement in *stability* of our sensory subtraction approach (please also refer to the video). On the other hand, comparing completion times and exerted forces during conditions $(K+C+P)_r$ and C_r lets us quantitatively analyze the loss of *performance* due to the absence of the kinesthetic component of the haptic interaction.

The intrinsic stability of cutaneous force feedback has been already discussed in Chap. 1, assessed in [7, 24–28], and here again highlighted by the success rates of the feedback conditions: 23 % for $(K+C)_r$ versus 100 % for C_r. As also clear from the video, the oscillations arose during condition $(K+C)_r$ made it very difficult for the subjects to complete the peg-in-hole task. On the other hand, employing cutaneous feedback alone (sensory subtraction) made the system stable, even without enforcing any stability control technique. Since the force is applied directly to the fingertips and does not affect the position of the end-effector of the master device, it is straightforward to assess that sensory subtraction would make *any* teleoperation system intrinsically stable (see Fig. 1.1b). However, it is also obvious that employing kinesthetic feedback in a stiff environment, without any stability control strategy, brought the system near to instability, making the trials difficult to complete. Nonetheless, in order to emphasize the stability properties of sensory subtraction, we decided to consider condition $(K+C)_r$ anyway.

The price for this improvement in stability is a significant loss of performance. In condition $(K+C+P)_r$, in fact, subjects completed the task in less time and exerted less force with respect to condition C_r. This means that full haptic feedback (if no oscillations arise) still leads to better performance with respect to employing cutaneous feedback alone. In Chap. 5 we will address this challenge and combine kinesthetic and cutaneous feedback to improve the performance of these teleoperation systems.

3.5 Conclusions

In this chapter we analyzed the feasibility and effectiveness of *sensory subtraction* in two challenging teleoperation scenarios. For this purpose, we developed a 3-DoF cutaneous display, able to apply cutaneous stimuli to the fingertips. In comparison to similar existing cutaneous devices, the one presented here has three actuated degrees of freedom, high peak force and accuracy, and it is able to provide the sensation of breaking and making contact with virtual and remote surfaces. Moreover, it can be easily attached to grounded haptic interfaces to provide, for example, additional kinesthetic feedback. We carried out two experiments of pinch grasp, both in virtual and real environments. This work extends the results presented in Chap. 2 and shows once more the feasibility of employing sensory subtraction in robotic teleoperation. Cutaneous feedback performed better than employing no force feedback at all, but, as expected, it was outperformed by full haptic feedback. However, sensory subtraction guaranteed the intrinsic stability of the teleoperation system and kept the system stable even in the absence of a stability controller. Kinesthetic feedback, on the other hand, showed highly degraded performance in such a case.

An improved version of the 3-DoF cutaneous device employed in this work will be used in the next chapter, where we will present an application of the sensory subtraction idea in a robotic surgery scenario.

References

1. L. Meli, C. Pacchierotti, D. Prattichizzo, Sensory subtraction in robot-assisted surgery: fingertip skin deformation feedback to ensure safety and improve transparency in bimanual haptic interaction. IEEE Trans. Biomed. Eng. **61**(4), 1318–1327 (2014)
2. B.J. Unger, A. Nicolaidis, P.J. Berkelman, A. Thompson, R.L. Klatzky, R.L. Hollis, Comparison of 3-D haptic peg-in-hole tasks in real and virtual environments, in *Proceedings of the IEEE/RJS International Conference on Intelligent Robots and Systems*, vol. 3 (2001), pp. 1751–1756
3. B. Hannaford, L. Wood, D.A. McAffee, H. Zak, Performance evaluation of a six-axis generalized force-reflecting teleoperator. IEEE Trans. Syst. Man Cybern. **21**(3), 620–633 (1991)
4. C. Pacchierotti, L. Meli, F. Chinello, M. Malvezzi, D. Prattichizzo, Cutaneous haptic feedback to ensure the stability of robotic teleoperation systems. Int. J. Robot. Res. (2015) http://ijr.sagepub.com/content/early/2015/10/15/0278364915603135.abstract (in press)
5. F. Chinello, M. Malvezzi, C. Pacchierotti, D. Prattichizzo, Design and development of a 3RRS wearable fingertip cutaneous device, in *Proceedings of the IEEE/ASME International Conference on Advanced Intelligent Mechatronics* (2015)
6. L. Meli, S. Scheggi, C. Pacchierotti, D. Prattichizzo, Wearable haptics and hand tracking via an RGB-D camera for immersive tactile experiences, in *Proceedings of the ACM Special Interest Group on Computer Graphics and Interactive Techniques Conference* (2014), p. 56
7. C. Pacchierotti, F. Chinello, M. Malvezzi, L. Meli, D. Prattichizzo, Two finger grasping simulation with cutaneous and kinesthetic force feedback. Haptics: Percept. Devices Mobil. Commun. **7282**, 373–382 (2012)
8. M.A. Srinivasan, K. Dankekar, An investigation of the mechanics of tactile sense using two dimensional models of the primate fingertip. Trans. ASME, J. Biomech. Eng. **118**, 48–55 (1996)
9. E.R. Serina, E. Mockensturm, C.D. Mote Jr, D. Rempel, A structural model of the forced compression of the fingertip pulp. J. Biomech. **31**(7), 639–646 (1998)
10. T. Cook, H. Alexander, M. Cohen, Experimental method for determining the 2-dimensional mechanical properties of living human skin. Med. Biol. Eng. Comput. **15**(4), 381–390 (1977)
11. M. Srinivasan, R.J. Gulati, K. Dandekar, In vivo compressibility of the human fingertip. ASME Bioeng. Decis. Publ. **22**, 573–576 (1992)
12. E.R. Serina, C.D. Mote et al., Force response of the fingertip pulp to repeated compression-effects of loading rate, loading angle and anthropometry. J. Biomech. **30**(10), 1035–1040 (1997)
13. N. Nakazawa, R. Ikeura, H. Inooka, Characteristics of human fingertips in the shearing direction. Biol. Cybern. **82**(3), 207–214 (2000). ISSN: 0340-1200
14. Q. Wang, V. Hayward, In vivo biomechanics of the fingerpad skin under local tangential traction. J. Biomech. **40**, 851–860 (2007)
15. M. Wiertlewski, V. Hayward, Mechanical behavior of the fingertip in the range of frequencies and displacements relevant to touch. J. Biomech. **45**(11), 1869–1874 (2012)
16. K. Salisbury, C. Tarr, Haptic rendering of surfaces defined by implicit functions, in *Proceedings of the ASME Dynamic Systems and Control Division*, vol. 61 (1997), pp. 61–67
17. M. Solazzi, W.R. Provancher, A. Frisoli, M. Bergamasco, Design of a SMA actuated 2-DoF tactile device for displaying tangential skin displacement, in *Proceedings of the IEEEWorld Haptics Conference* (2011), pp. 31–36

18. Z.F. Quek, S.B. Schorr, I. Nisky, A.M. Okamura, W.R. Provancher, Sensory augmentation of stiffness using fingerpad skin stretch, in *Proceedings of the World Haptics Conference* (2013), pp. 467–472
19. K.H. Park, B.H. Kim, S. Hirai, Development of a soft-fingertip and its modeling based on force distribution, in *Proceedings of the of IEEE International Conference on Robotics and Automation*, vol. 3 (2003), pp. 3169–3174
20. C.B. Zilles, J.K. Salisbury, A constraint-based god-object method for haptic display, in *Proceedings of IEEE/RSJ International Conference on Intelligent Robots and Systems*, vol. 3 (1995), pp. 146–151
21. M.J. Massimino, T.B. Sheridan, Sensory substitution for force feedback in teleoperation. Presence: Teleoperators Virtual Environ. **2**(4), 344–352 (1993)
22. D. Prattichizzo, J. Trinkle, Chapter 28 on grasping, in *Handbook on Robotics* (Springer, Berlin, 2008), pp. 671–700
23. M. Franken, S. Stramigioli, S. Misra, C. Secchi, A. Macchelli, Bilateral telemanipulation with time delays: a two-layer approach combining passivity and transparency. IEEE Trans. Robot. **27**(4), 741–756 (2011)
24. D. Prattichizzo, C. Pacchierotti, G. Rosati, Cutaneous force feedback as a sensory subtraction technique in haptics. IEEE Trans. Haptics **5**(4), 289–300 (2012)
25. C. Pacchierotti, A. Tirmizi, D. Prattichizzo, Improving transparency in teleoperation by means of cutaneous tactile force feedback. ACM Trans. Appl. Percept. **11**(1), 4:1–4:16 (2014)
26. C. Pacchierotti, F. Chinello, D. Prattichizzo, Cutaneous device for teleoperated needle insertion, in *Proceedings of the 4th IEEE RAS EMBS International Conference on Biomedical Robotics and Biomechatronics (BioRob)* (2012), pp. 32–37
27. C. Pacchierotti, D. Prattichizzo, K.J. Kuchenbecker, Cutaneous feedback of fingertip deformation and vibration for palpation in robotic surgery, IEEE Trans. Biomed. Eng. in press (2015)
28. C. Pacchierotti, A. Tirmizi, G. Bianchini, D. Prattichizzo, Enhancing the performance of passive teleoperation systems via cutaneous feedback, IEEE Trans. Haptics in press (2015)

Chapter 4
Remote Palpation Using the da Vinci Surgical System

Abstract The main advantage of using cutaneous-only feedback techniques, with respect to more popular kinesthetic approaches, is a higher degree of safety. In fact, cutaneous feedback guarantees the intrinsic stability of the teleoperation system. This is especially promising for those scenarios where safety is a paramount and non-negotiable requirement, such as in robot-assisted surgery. This chapter extends the previous evaluations of sensory subtraction to a robotic surgery scenario. It presents a novel force feedback system for the da Vinci Surgical System, capable of providing cutaneous feedback to the surgeon while guaranteeing the safety of the patient. Designed to provide contact deformation and vibration cues, the system is composed of a BioTac tactile sensor, mounted to one of the robot's slave tools, and a cutaneous device, attached to the corresponding master controller. Contact deformations and vibrations sensed by the BioTac are directly mapped to input commands for the cutaneous device's motors using a novel model-free data-driven algorithm. The cutaneous display continually moves, tilts, and vibrates a flat plate at the operator's fingertip to optimally reproduce the tactile sensations experienced by the BioTac. The proposed approach was tested by having eighteen subjects use the augmented da Vinci robot to palpate a heart model. Cutaneous feedback significantly improved palpation performance in all the considered metrics.

4.1 Introduction

Chapter 2 presented an application of our sensory subtraction idea in a simple 1-DoF simulated needle insertion scenario. Chapter 3 extended the evaluation to a more challenging peg-in-hole task, carried out both in virtual and real environments. This chapter extends the results presented until now to a surgical scenario, which is the natural application of our sensory subtraction technique. As discussed in Chap. 1 and [1–5], in fact, sensory subtraction guarantees the intrinsic stability of teleoperation systems, and it is thus very promising for those scenarios where safety is a paramount

---

This chapter is reprinted with kind permission from IEEE, originally published in [6, 7].

C. Pacchierotti, *Cutaneous Haptic Feedback in Robotic Teleoperation*,
Springer Series on Touch and Haptic Systems, DOI 10.1007/978-3-319-25457-9_4

and non-negotiable requirement, such as robot-assisted surgery. Section 1.3 provides an overview on cutaneous technologies used for robot-assisted surgery.

This chapter presents a novel cutaneous feedback system for the da Vinci surgical robot. It is designed to provide contact deformation and vibrotactile feedback to the surgeon while guaranteeing the stability and safety of the teleoperation system. It is composed of a BioTac tactile sensor, mounted to one of the robot's slave tools, and a cutaneous display, attached to the corresponding master controller. Contact deformations and vibrations sensed by the BioTac are directly mapped to input commands for the cutaneous device's motors using a novel model-free data-driven algorithm.

Section 4.2 presents and evaluates the algorithm considering only contact deformations. Section 4.3 extends the approach to include also vibrotactile cues. Finally, Sect. 4.4 addresses concluding remarks and perspectives of the work.

4.2 Cutaneous Feedback System for Remote Interaction: Contact Deformations

This section presents the tactile rendering algorithm for contact deformations. It directly maps the sensed stimuli to the best possible input commands for the device's motors using a dataset recorded with the tactile sensor inside the device. We considered a haptic system composed of a BioTac tactile sensor, in charge of measuring contact deformations, and a custom 3-DoF cutaneous device with a flat contact platform, in charge of applying deformations to the user's fingertip. To validate the proposed approach and discover its inherent trade-offs, we carried out two remote tactile interaction experiments. The first evaluated the error between the tactile sensations registered by the BioTac in a remote environment and the sensations created by the cutaneous device for six representative tactile interactions and 27 variations of the display algorithm. The average errors in the best condition were 3.0 % of the BioTac's full 12-bit scale. The second experiment evaluated human subjects' experiences for the same six remote interactions and eight algorithm variations. The average subjective rating for the best algorithm variation was 8.2 out of 10, where 10 is best.

Section 4.3 will extend this approach to include also vibrotactile cues and will validate the cutaneous system in a challenging palpation task using the da Vinci surgical robot.

4.2.1 Sensing and Actuation Systems

The BioTac tactile sensor mimics the physical properties and sensory capabilities of the human fingertip [8, 9]. As shown in Fig. 4.1a, it consists of three complementary

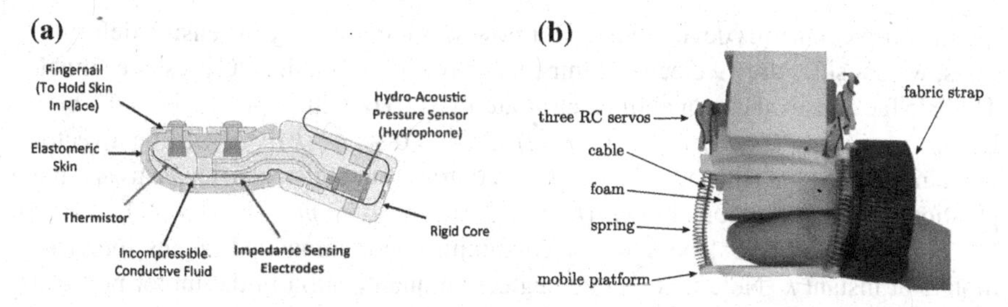

Fig. 4.1 The haptic system considered in this work includes a BioTac tactile sensor, in charge of registering contact deformations at the remote environment, and a custom 3-DoF cutaneous device, in charge of applying those deformations to the user's fingertip through a mobile platform controlled by three RC servos. **a** The BioTac sensor (picture courtesy of SynTouch LLC). **b** The 3-DoF cutaneous device

sensory systems (deformation, internal fluid pressure, and temperature) integrated into a single package. Contact forces deform the elastic skin and the underlying conductive fluid, changing the impedances of 19 electrodes distributed over the surface of the rigid core. The DC pressure of the conductive fluid is measured by a hydro acoustic pressure sensor, which also detects the AC pressure changes caused by transient contacts such as textures. The BioTac is internally heated to near human body temperature, and both DC and AC temperature are measured by a thermistor placed near the surface of the rigid core.

The haptic interface employed here is a 3-DoF fingertip cutaneous device that is similar to the one presented in Chap. 3 and in [2, 10, 11]. As shown in Fig. 4.1b, it is composed of a static platform that houses three servo motors above the user's fingernail and a mobile platform that applies the requested stimuli to the fingertip. Three cables connect the two platforms, and springs around the cables keep the mobile platform in a reference configuration, away from the fingertip, when not actuated. By controlling the cable lengths, the motors can orient and translate the mobile platform in three-dimensional space. The device fastens to the finger with a fabric strap. The servo motors used in our prototype are Sub-Micro Servo 3.7 g motors (Pololu Corporation, Las Vegas, USA), which are able to exert up to 39 N·mm torque and have a positioning resolution of 0.5°. A short video showing the device is available as supplemental material at http://extras.springer.com/978-3-319-25455-5 and at http://goo.gl/oVw56j.

4.2.2 Mapping Between Remote Sensed Data and Motor Commands

Our goal is to enable the user to perceive, through the 3-DoF fingertip cutaneous device, the deformations experienced by the BioTac in the remote environment. In other words, we aim to find an effective many–to–few mapping between the rich sensory information provided by the BioTac and the limited actuation capabilities of the fingertip cutaneous device. Since, for now, our focus is only in sensing deformations, we consider the 19 electrode impedance readings and the DC pressure signal. The BioTac senses these quantities at a rate of 100 Hz with a precision of 12 bits. Let $\boldsymbol{s_d(k)} \in \mathbb{S}_d = \{(s_{d,1}(k), \ldots, s_{d,20}(k)) \in \mathbb{Z}^{20} : 0 \le s_{d,i}(k) \le 4095\}$ be a vector containing the values sensed at instant k. In contrast, our cutaneous device uses three position-controlled motors. Let $\boldsymbol{m_d(k)} \in \mathbb{M}_d = \{(m_{d,1}(k), m_{d,2}(k), m_{d,3}(k)) \in \mathbb{R}^3 : 30^\circ \le m_{d,i}(k) < 195^\circ\}$ be a vector containing the commanded angles for these motors at instant k. Note that we are neglecting quantization in the motor position outputs for simplicity. In order to simplify the notation further, the sampling time index k will be omitted from now on.

How can we map a given BioTac sensation to a congruent configuration of the mobile platform? Because the BioTac mimics the *physical* properties of the human fingertip [8, 9], we realized that we could place it inside the cutaneous device to discover how the motion of the mobile platform affects the tactile sensor. As shown in Fig. 4.2, the BioTac was placed between the foam and the mobile platform, in the same way a human user would wear the device. We then moved the mobile platform to a wide range of configurations and registered the effect of each of these configurations on the BioTac, saving both the commanded motor angles $\boldsymbol{m_{d,*}}$ and the resulting effect on the tactile sensor $\boldsymbol{s_{d,*}}$. Using a moderate step size of $\theta = 3^\circ$ yields $\left(\frac{195^\circ - 30^\circ}{3^\circ}\right)^3 = 1{,}66{,}375$ unique platform configurations. The platform was held in each configuration for 0.1 s, and the values gathered by the BioTac were arithmetically averaged. Data collection took approximately 47 h. At the end of data collection, we were thus able to evaluate the mapping function

$$\begin{aligned} \mu_d &: \mathbb{S}_{d,*} \to \mathbb{M}_{d,*}, \\ \mu_d(\boldsymbol{s_{d,*}}) &= \boldsymbol{m_{d,*}}, \end{aligned} \tag{4.1}$$

which links the BioTac sensed data to the motor input commands. Set $\mathbb{M}_{d,*} \subset \mathbb{M}_d$ contains all the angle triplets actuated during data collection, and $\mathbb{S}_{d,*} \subset \mathbb{S}_d$ contains all the resulting sensed values registered by the BioTac. Because we have only one corresponding point in $\mathbb{S}_{d,*}$ for each point in $\mathbb{M}_{d,*}$, and vice versa, function $\mu_d(\cdot)$ is bijective. In this case the cardinality of sets $\mathbb{S}_{d,*}$ and $\mathbb{M}_{d,*}$ is 1,66,375, which is much lower than the 4096^{20} different points the BioTac can sense (i.e., $|\mathbb{S}| = 4096^{20}$).

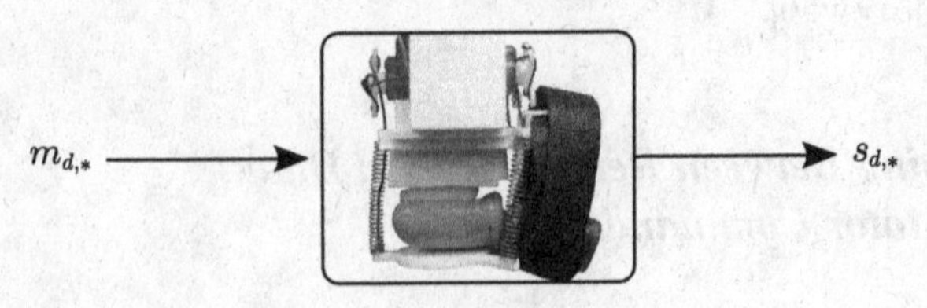

Fig. 4.2 Data collection. The BioTac was placed inside the cutaneous device, and the platform was moved to a wide range of configurations. The motor inputs $\boldsymbol{m_{d,*}}$ and the resultant cutaneous sensations $\boldsymbol{s_{d,*}}$ were recorded

Function $\mu_d(\cdot)$ is thus defined for a very small subset of all the possible tactile sensations the BioTac can experience. For this reason, we cannot simply deploy the sensor in a remote environment and expect its sensed points to be in the domain of our mapping function $\mu_d(\cdot)$. Unfortunately, this problem cannot be fixed by simply reducing the angle step size during data collection. The shape of the platform and the limited degrees of freedom of the cutaneous device will always couple the behavior of neighboring electrodes, so not all points in $\mathbb{S}_d$ are reachable with the given device. Other low-DoF cutaneous devices would be able to reach different subsets of $\mathbb{S}_d$.

We thus need a function that maps a *generic* sensed point $\boldsymbol{s_d} \in \mathbb{S}_d$ to one in our mapping function's domain $\mathbb{S}_{d,*}$. An idea, for example, is to look for the point in our domain closest to the sensed one, thus defining

$$\begin{aligned} \nu_d &: \mathbb{S}_d \rightarrow \mathbb{S}_{d,*}, \\ \nu_d(\boldsymbol{s_d}) &= \boldsymbol{s_{d,*}}, \end{aligned} \tag{4.2}$$

as the function that maps a generic point $\boldsymbol{s_d} \in \mathbb{S}_d$, sensed by the BioTac, to the closest one in $\mathbb{S}_{d,*}$. In this work we implemented the nearest point search using the Approximate Nearest Neighbour (ANN) C++ library by Mount and Arya [12], using the 20-dimensional Euclidean distance metric. In order to evenly weight the twenty elements of the sensed data when computing the distance, we divided each component of $\boldsymbol{s_d}$ and $\boldsymbol{s_{d,*}}$ by the corresponding standard deviation observed during data collection, so that the standard deviation of each component of the vectors in $\mathbb{S}_{d,*}$ becomes 1.

It is now trivial to combine functions $\mu_d(\cdot)$ and $\nu_d(\cdot)$ to define

$$\begin{aligned} f_d &: \mathbb{S}_d \rightarrow \mathbb{M}_{d,*}, \\ f_d(\boldsymbol{s_d}) &= \mu_d(\nu_d(\boldsymbol{s_d})) = \mu_d(\boldsymbol{s_{d,*}}) = \boldsymbol{m_{d,*}}, \end{aligned} \tag{4.3}$$

as our final function, which maps a generic point $\boldsymbol{s_d} \in \mathbb{S}_d$, sensed by the BioTac, to the motor angle triplet $\boldsymbol{m_{d,*}} \in \mathbb{M}_{d,*}$ that most closely causes sensation $\boldsymbol{s_d}$. Since $\mu_d(\cdot)$ is a bijective function and $\nu_d(\cdot)$ a surjective function, their composition $f_d(\cdot)$ is a surjection.

Although $f_d(\cdot)$ provides an effective way to map a generic point sensed by the BioTac to a motor angle triplet, its image set $\mathbb{M}_{d,*}$ contains a very small subset of all the possible angle configurations the motors can reach. Our 3° step size in data collection yielded only $\frac{195^\circ-30^\circ}{3^\circ} = 54$ different angle configurations out of the $\frac{195^\circ-30^\circ}{0.5^\circ} = 330$ configurations that each motor can reach. This problem can be easily addressed by choosing a smaller step size. However, considering that the present data collection took 47 hours, this approach may not always be feasible. A step size of $\theta = 0.5^\circ$ (the servo angle resolution) would yield $\left(\frac{195^\circ-30^\circ}{0.5^\circ}\right)^3 \approx 3.6 \cdot 10^7$ unique platform configurations and an estimated duration of more than one year for data collection, which is not justified. Moreover, having so many points in $\mathbb{S}_{d,*}$ would also impose severe computational constraints on the algorithm speed.

An alternative way to address this problem is redefining our search function $\nu_d(\cdot)$ to provide more than one neighbor of the point sensed by the BioTac. The resulting motor angle triplets can then be combined using a weighted average. In addition to enlarging the image set of our final function, this approach also makes the system more robust to isolated errors during data collection. Let us thus define

$$\nu_{d,n} : \mathbb{S}_d \to \mathbb{S}^n_{d,*},$$
$$\nu_{d,n}(\boldsymbol{s_d}) = \begin{bmatrix} \boldsymbol{s}_{d,*,1} \\ \boldsymbol{s}_{d,*,2} \\ \vdots \\ \boldsymbol{s}_{d,*,n} \end{bmatrix} = \widehat{\boldsymbol{s}}_{d,*}, \tag{4.4}$$

as our improved search function that maps a generic point $\boldsymbol{s_d} \in \mathbb{S}_d$, sensed by the BioTac, to the n closest ones in $\mathbb{S}_{d,*}$, and

$$\mu_{d,n} : \mathbb{S}^n_{d,*} \to \mathbb{M}^n_{d,*},$$
$$\mu_{d,n}(\widehat{\boldsymbol{s}}_{d,*}) = \mu_{d,n}\left(\begin{bmatrix} \boldsymbol{s}_{d,*,1} \\ \boldsymbol{s}_{d,*,2} \\ \vdots \\ \boldsymbol{s}_{*,n} \end{bmatrix}\right) = \begin{bmatrix} \mu(\boldsymbol{s}_{d,*,1}) \\ \mu(\boldsymbol{s}_{d,*,2}) \\ \vdots \\ \mu(\boldsymbol{s}_{d,*,n}) \end{bmatrix}$$
$$= \begin{bmatrix} \boldsymbol{m}_{d,*,1} \\ \boldsymbol{m}_{d,*,2} \\ \vdots \\ \boldsymbol{m}_{d,*,n} \end{bmatrix} = \widehat{\boldsymbol{m}}_{d,*}, \tag{4.5}$$

as the function that maps those n points to their corresponding motor angle triplets registered during data collection.

We now require an additional step to map $\widehat{\boldsymbol{m}}_{d,*} \in \mathbb{M}^n_{d,*}$ to a single angle triplet for the servo motors. Let us thus define

$$\varphi_{d,n} : \mathbb{M}^n_{d,*} \to \mathbb{M}_d,$$
$$\varphi_{d,n}(\widehat{\boldsymbol{m}}_{d,*}) = \varphi_{d,n}\left(\begin{bmatrix} \boldsymbol{m}_{d,*,1} \\ \boldsymbol{m}_{d,*,2} \\ \vdots \\ \boldsymbol{m}_{d,*,n} \end{bmatrix}\right) = \overline{\boldsymbol{m}}_d, \tag{4.6}$$

as the function that averages the n angle triplets in $\widehat{\boldsymbol{m}}_{d,*}$ to yield a single one defined in the set of *all* the angle triplets reachable by our motors. This work used a simple inverse-distance weighted mean. We thus computed each component of $\overline{\boldsymbol{m}}_d = (\overline{m}_{d,1}, \overline{m}_{d,2}, \overline{m}_{d,3})$ as

$$\overline{m}_{d,i} = \sum_{q=1}^{n} \frac{w_q \cdot m_{d,*,q,i}}{\sum_{p=1}^{n} w_p}, \quad i = 1, 2, 3, \tag{4.7}$$

where $m_{d,*,q,i}$ is the ith component of $\boldsymbol{m}_{d,*,q}$, and $w_q = (|\boldsymbol{s}_d - \boldsymbol{s}_{d,*,q}|)^{-2}$ is the reciprocal of the square of the distance between the sensed point $\boldsymbol{s}_d$ and the qth point closest to it in $\mathbb{S}_{d,*}$, as evaluated by $\nu_{d,n}(\cdot)$.

Finally, we can combine $\mu_{d,n}(\cdot)$, $\nu_{d,n}(\cdot)$ and $\varphi_{d,n}(\cdot)$ to define

$$\begin{aligned} f_{d,n} &: \mathbb{S}_d \rightarrow \mathbb{M}_d, \\ f_{d,n}(\boldsymbol{s}_d) &= \varphi_{d,n}(\mu_{d,n}(\nu_{d,n}(\boldsymbol{s}_d))) = \varphi_{d,n}(\mu_{d,n}(\widehat{\boldsymbol{s}}_{d,*})) \\ &= \varphi_{d,n}(\widehat{\boldsymbol{m}}_{d,*}) = \overline{\boldsymbol{m}}_d, \end{aligned} \tag{4.8}$$

as our enhanced final function, which maps a generic point $\boldsymbol{s}_d \in \mathbb{S}_d$, sensed by the BioTac, to a motor angle triplet $\overline{\boldsymbol{m}}_d \in \mathbb{M}_d$. It is worth noting that $\mu_d(\cdot)$ and $\nu_d(\cdot)$, defined respectively in Eqs. (4.1) and (4.2), are particular cases of $\mu_{d,n}(\cdot)$ and $\nu_{d,n}(\cdot)$ for $n = 1$. Moreover, function $\varphi_{d,n}(\cdot)$ for $n = 1$ is the identity function, so $f_d(\cdot)$ defined in Eq. (4.3) is also a particular case of $f_{d,n}(\cdot)$ for $n = 1$.

The algorithm is summarized in Fig. 4.3. Its performance is expected to improve when one reduces the angle step size in data collection, increases the degrees of freedom of the device, and/or increases the number of neighbors retrieved by function $\nu_{d,n}(\cdot)$. A short video featuring the BioTac interacting with a remote environment and the cutaneous device driven according to this algorithm is available as supplemental material at http://extras.springer.com/978-3-319-25455-5 and at http://goo.gl/GAuGDp. The system shown in the video uses only one neighbor, as in Eq. (4.3) and [13].

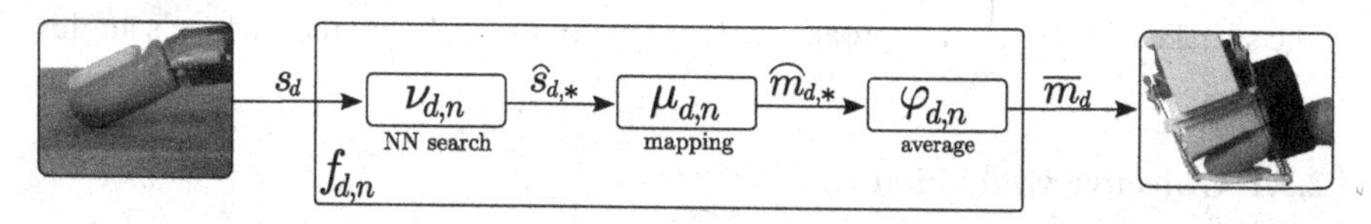

Fig. 4.3 Rendering algorithm for contact deformations only. The BioTac registers a tactile sensation $\boldsymbol{s}_d$ at the remote environment. Function $\nu_{d,n}(\cdot)$ looks for the n closest points $\boldsymbol{s}_{d,*}$ contained in the set recorded during data collection. These points are then mapped by $\mu_{d,n}(\cdot)$ to the corresponding motor angle triplets $\boldsymbol{m}_{d,*}$. Finally, $\varphi_{d,n}(\cdot)$ averages those points to find the angle triplet $\overline{\boldsymbol{m}}_d$ that should be actuated by the device to allow the user to feel an approximation of what the BioTac is feeling. This algorithm will be extended to consider also vibrotactile cues in Sect. 4.3

4.2.3 Experimental Evaluation

We evaluated the proposed algorithm by carrying out two remote tactile interaction experiments. The first experiment aims to quantitatively evaluate the error between the tactile sensations registered by the BioTac in the remote environment and the ones actuated by the cutaneous device. The second experiment aims to collect quantitative data regarding the experience of human subjects using our tactile system.

To enable us to compare their results, the two experiments used the same set of remote tactile experiences. We recorded video and tactile data during six different interactions between a stationary BioTac sensor and a flat metal surface that was moved by hand to touch the BioTac's fingertip in different ways. The experimental setup is shown in Fig. 4.4. A short video featuring all the interactions is available as supplemental material at http://extras.springer.com/978-3-319-25455-5 and at http://goo.gl/PsuDTV. The first interaction, called *back-flat-back* in Fig. 4.4a, consists of making contact with the proximal part of the BioTac, then moving toward the finger pulp and returning to the back part of the finger. The second one, called *tip-flat-tip* in Fig. 4.4b, consists of making contact with the distal part of the BioTac, then moving toward the finger pulp and returning to the tip. The third one, called *left-flat-left* in Fig. 4.4c, consists of making contact on the left lateral side of the BioTac, then moving toward the finger pulp and returning to the left side. The fourth one, called *right-flat-right* in Fig. 4.4d, performs the same sequence of interactions on the right lateral side of the BioTac. The fifth one, called *flat* in Fig. 4.4e, consists of making contact with the finger pulp with the surface parallel to the sensor's nail. The last one, called *complex* in Fig. 4.4f, is a combination of the other five motions. Each interaction was repeated three times within its recording. The whole tactile experience (six interactions, each of them repeated three times) lasts 3.4 min. Figure 4.5 shows representative frames of the videos for the *back-flat-back* and *tip-flat-tip* interactions, together with the respective platform configurations chosen by the proposed algorithm. Note that the cutaneous device's platform mimics the relative orientation of the surface although our approach does not explicitly measure the surface's angle.

4.2.3.1 Objective Evaluation

In order to evaluate the error between the tactile sensations registered in the remote environment and the ones actuated by the cutaneous device, we placed a BioTac sensor inside the cutaneous device, as done for data collection (see Fig. 4.2). We then sent the tactile data recorded during the six interactions of Fig. 4.4 through our algorithm and drove the cutaneous device to the resulting motor angle triplets over time, while recording what the BioTac sensed.

We compared the results of nine different versions of the algorithm's parameters, varying the step size during data collection ($\theta = 3°, 6°, 9°$) and the number of points retrieved by the nearest neighbor function $\nu_{d,n}(\cdot)$, $n = 1, 4, 8$ (see Eq. 4.4). Moreover, to show the generality of the algorithm, we also considered three different

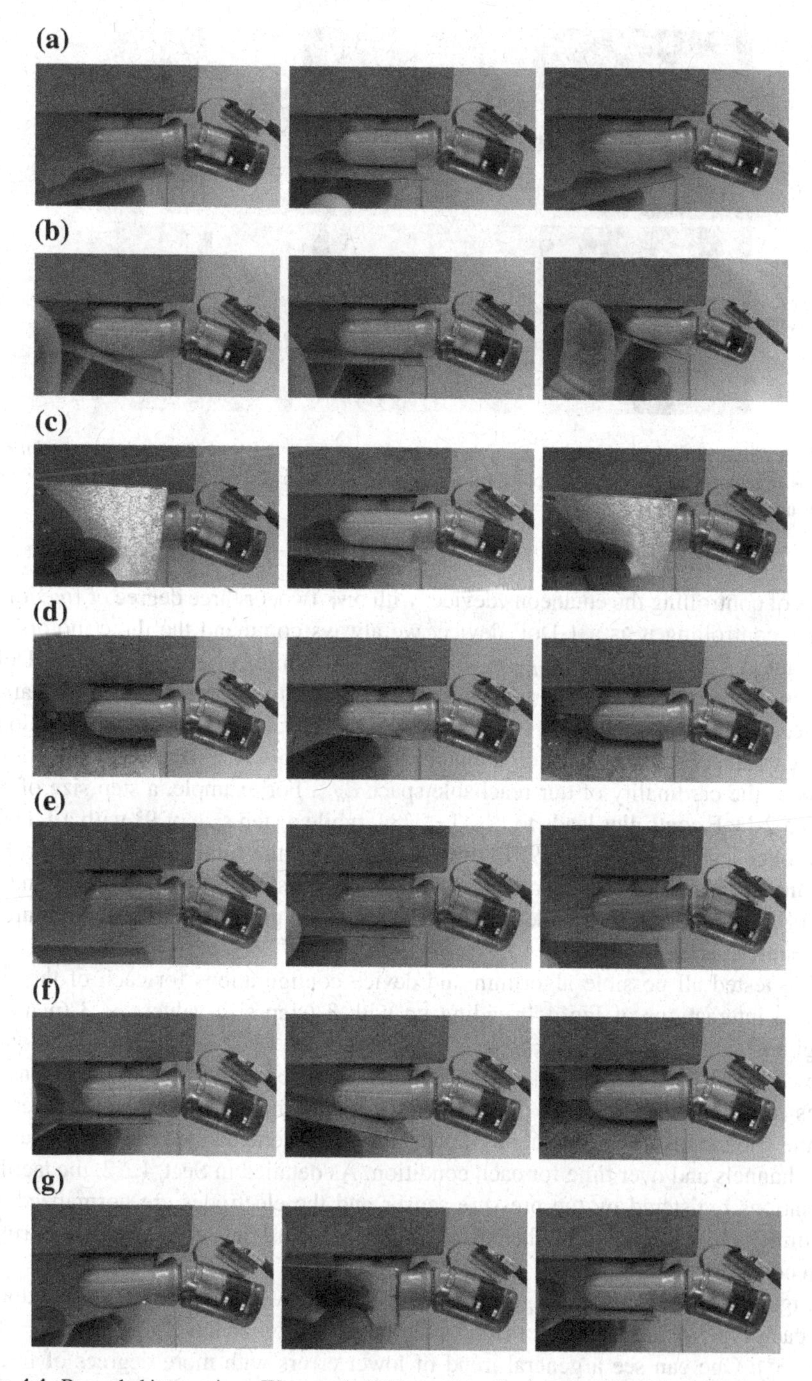

Fig. 4.4 Recorded interactions. We recorded video and tactile data during six different interactions between a BioTac sensor and a flat surface. **a** Back-flat-back interaction. **b** Tip-flat-tip interaction. **c** Left-flat-left interaction. **d** Right-flat-right interaction. **e** Flat interaction. **g** Complex interaction (a combination of the others)

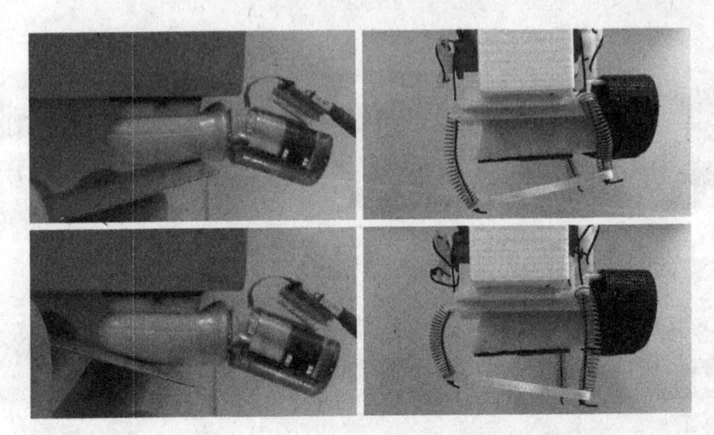

Fig. 4.5 Algorithm demonstration. Representative frames from the *back-flat-back* and *tip-flat-tip* videos, together with the respective platform configurations as chosen by the algorithm. The depicted cases used a step size of 3° and $n = 8$ neighbors

ways of controlling the cutaneous device: with one, two, or three degree of freedom. When controlling it as a 1-DoF device, we always command the three motors to the same target angle (i.e., $m_{d,1} = m_{d,2} = m_{d,3}$). While controlling it as a 2-DoF device, we give the two rear motors (the ones closer to the fabric strap) the same angle, and the front motor is driven independently. Reducing the degrees of freedom of the cutaneous device, as well as increasing the step size during data collection, reduces the cardinality of our reachable space $\mathbb{S}_{d,*}$. For example, a step size of 6° with a 2-DoF controller leads to $|\mathbb{S}_{d,*}| = 784$, while a step size of 9° with a 1-DoF controller leads to $|\mathbb{S}_{d,*}| = 19$. In order to guarantee the same setup for all of the conditions, we ran the data collection only once, considering a 3° step size and a 3-DoF configuration for the device, and from that we generated all of the required reachable spaces.

We tested all possible algorithm and device configurations for each of the six remote interactions of Fig. 4.4, ending up with 3 (step size values) × 3 (nearest neighbor values) × 3 (DoF values) × 6 (interactions) = 162 different conditions. The experiment lasted 96 min. We calculated the error between the tactile sensations registered by the BioTac in the remote environment and the ones registered by the BioTac inside the cutaneous device and then averaged the error across sensing channels and over time for each condition. As detailed in Sect. 4.2.2, the tactile sensations registered by the pressure sensor and the electrodes are normalized at runtime by dividing them by the corresponding standard deviation observed during data collection.

Figure 4.6 depicts the average normalized error versus device degrees of freedom for each type of interaction for a representative version of the algorithm ($n = 1$, $\theta = 3°$). One can see a general trend of lower errors with more degrees of freedom. Furthermore, the simplest interaction (*flat*) has approximately the same error for all versions of the device controller, while the motions that touch the sides of

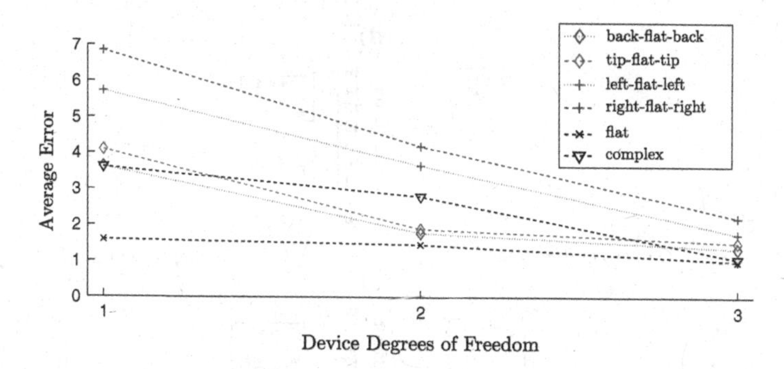

Fig. 4.6 Selected results from objective evaluation. Normalized error versus degrees of freedom for each tactile interaction for a representative version of the algorithm ($n = 1$, $\theta = 3°$). While separated in this plot, the six tactile interactions were considered independent observations in the data analysis

the BioTac have the highest errors and are most sensitive to controller version. Each datapoint in this plot represents just a single measurement, so we averaged the errors across the six interactions to enable statistical analysis. Figure 4.7 shows the results averaged across the six tactile interactions. It is also useful to consider the average non-normalized error with respect to the full 12-bit scale the BioTac can reach (0–4095). The average error in the best condition (3-DoF, $\theta = 3°$, $n = 1$) was 3.0 % of the full 12-bit scale, and it was 31.6 % of full scale in the worst condition (1-DoF, $\theta = 9°$, $n = 1$).

To compare the different algorithm and device configurations, we ran a three-way repeated-measures ANOVA on the normalized error data shown in Fig. 4.7. Each tactile interaction was considered as an independent observation. Step size during data collection, number of neighbors retrieved by function $\nu_{d,n}(\cdot)$, and degrees of freedom of the device were treated as within-subject factors.

All the data passed the Shapiro-Wilk normality test. Mauchly's Test of Sphericity indicated that the assumption of sphericity had not been violated for the device degrees of freedom, while it was violated for the data collection step size ($\chi^2(2) = 17.008$, $p < 0.001$), number of neighbors retrieved ($\chi^2(2) = 21.705$, $p < 0.001$), interaction between degrees of freedom and step size ($\chi^2(9) = 40.938$, $p < 0.001$), interaction between degrees of freedom and number of neighbors retrieved ($\chi^2(9) = 38.852$, $p < 0.001$), and interaction between step size and number of neighbors retrieved ($\chi^2(9) = 44.385$, $p < 0.001$). A Greenhouse-Geisser correction was applied to the tests involving data that violate the sphericity assumption.

The ANOVA test revealed a statistically significant change in the error for degrees of freedom ($F(2, 10) = 31.415$, $p < 0.001$), step size ($F(1.007, 5.036) = 43.791$, $p = 0.001$), and number of neighbors retrieved ($F(1.002, 5.011) = 35.943$, $p = 0.002$). Moreover, there was a statistically significant interaction between degrees of freedom and step size ($F(1.146, 5.728) = 30.785$, $p = 0.001$), degrees of freedom and number of neighbors retrieved ($F(1.235, 6.177) = 25.528$, $p = 0.002$), and step size and number of neighbors retrieved ($F(1.016, 5.082) = 43.929$, $p = 0.001$).

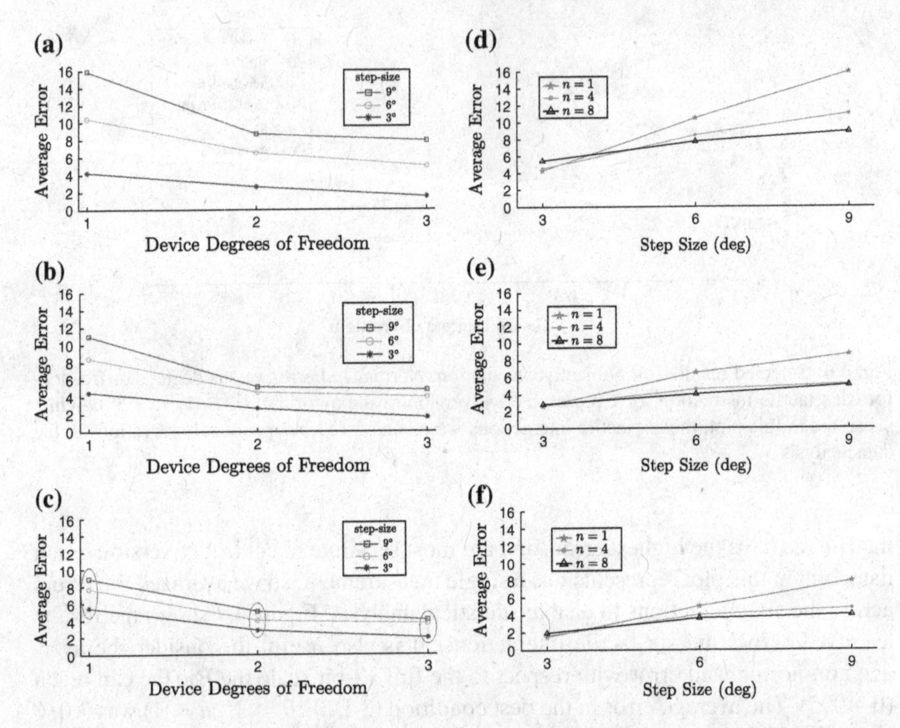

Fig. 4.7 Results of objective evaluation averaged across tactile interactions. Average normalized error versus degrees of freedom (**a**, **b**, **c**) and step size (**d**, **e**, **f**) are reported. Statistical analysis revealed a significant increase in the rendering error when reducing the degrees of freedom of the cutaneous device, when increasing the step size during data collection, and when reducing the number of neighbors retrieved by function $\nu_{d,n}(\cdot)$. **a** Degrees of freedom versus error, n = 1. **b** Step size versus error, 1-DoF configuration. **c** Degrees of freedom versus error, n = 4. **d** Step size versus error, 2-DoF configuration. **e** Degrees of freedom versus error, n = 8. **f** Step size versus error, 3-DoF configuration

Post hoc analysis with Bonferroni adjustments revealed a significant increase in the rendering error when reducing the degrees of freedom of the cutaneous device (1-DoF vs. 2-DoF, $p = 0.013$; 1-DoF vs. 3-DoF, $p = 0.002$; 2-DoF vs. 3-DoF, $p = 0.036$), when increasing the step size during data collection (3° vs. 6°, $p = 0.005$; 3° vs. 9°, $p = 0.003$; 6° vs. 3°, $p = 0.002$), and when reducing the number of neighbors retrieved by function $\nu_{d,n}(\cdot)$ (1 vs. 4, $p = 0.006$; 1 vs. 8, $p = 0.0005$; 4 vs. 8, $p = 0.013$).

As it is clear from Fig. 4.7 and from the statistical analysis reported above, there is a significant interaction between step size and number of neighbors retrieved by the algorithm: increasing the number of neighbors retrieved reduces the difference between conditions with dissimilar step sizes. To determine whether this difference can be considered statistically negligible, we used the two one-sided t-test approach (TOST) [3, 14, 15]. The null hypothesis of the TOST states that the mean values of two groups are different by at least a certain amount ε. Then, in order to test for equivalence, the 90 % confidence intervals for the difference between the two groups are evaluated. The null hypothesis that the groups differ by at least ε is rejected if the limits of the interval fall outside the $\pm\varepsilon$ bounds. Conversely, comparability is demonstrated when the bounds of the 90 % confidence interval of the mean difference fall entirely within the $\pm\varepsilon$ bounds [14, 16]. The design of equivalence tests can be tricky because the acceptance criterion ε must be defined on the basis of prior knowledge of the measurement. For a sample data set of p independent measurements with standard deviation δ, for instance, ε must certainly be greater than $\delta/\sqrt{p}$, otherwise the test may fail simply because of imprecision, rather than because of a true difference. However, it must also be less than any specifications or standards that the testing is challenging, or the test becomes too easy and will not adequately discriminate.

In this work we evaluated ε as suggested in [14], where the authors provided a useful step-by-step process for performing equivalence testing with commonly available computational software packages. The two one-sided tests were performed between conditions retrieving $n = 8$ neighbors, considering separately conditions with different degrees of freedom. The three tested groups are circled in Fig. 4.7c. To avoid raising the family-wise error rate, i.e., the probability of at least one incorrectly rejected null hypothesis in a family of tests, we took into account the simple correction discussed in [17]. The tests revealed statistical equivalence between all three step sizes (3°, 6° and 9°) when actuating one degree of freedom, and between 3° and 6° when actuating two and three degrees of freedom.

4.2.3.2 Subjective Evaluation

We carried out a second experiment to evaluate the subjective experience of using the presented tactile system. We considered the same six tactile interactions used in Sect. 4.2.3.1, but we had to reduce the number of algorithm and device versions to keep the experiment to a reasonable duration for human subjects. Thus, we compared the two extreme step sizes (3° and 9°), the two extreme functions $\nu_{d,n}(\cdot)$ (retrieving 1 and 8 neighbors), and the two extreme ways of controlling the cutaneous device (one and three degrees of freedom). Similar to Sect. 4.2.3.1, we tested these conditions for each of the six remote interactions, ending up with 2 (step size values) $\times$ 2 (nearest neighbor valuess) $\times$ 2 (DoF values) $\times$ 6 (interactions) = 48 different conditions.

As shown in Fig. 4.8, the subject was asked to wear the cutaneous device on his or her right index finger and look at a 61-cm-diagonal LCD screen that presented the six tactile interaction videos shown in Fig. 4.4. Each video was played eight times, once for each considered combination of the algorithm parameters and the device controller. As the video played, the system sent the corresponding recorded tactile data through our algorithm and drove the cutaneous device accordingly, so that the subject felt on his or her finger a particular rendering of what the BioTac was experiencing in the video. Subjects were isolated from external noise through a pair of headphones playing white noise, and their vision of the cutaneous device was blocked by a cardboard panel. At the end of each video, the subject was asked to rate how well the cutaneous device replicated the sensations experienced by the

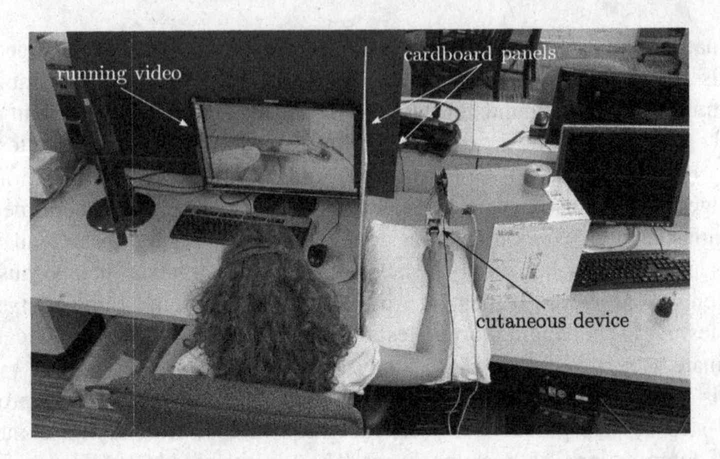

Fig. 4.8 Experimental setup for subjective evaluation. The subject was asked to wear the cutaneous device on his or her right index finger and watch videos featuring the six tactile interactions. At the end of each video, the subject rated how well the cutaneous device replicated the sensations experienced by the sensor in the video

sensor in the video. The response was given using a slider that ranged from 0 to 10, where a score of 0 meant "very badly" and a score of 10 meant "very well".

To determine the number of subjects needed for our research study, we ran a power analysis using the G*Power software. We estimated the effect size from the data retrieved in Sect. 4.2.3.1, expecting conditions showing higher errors in Sect. 4.2.3.1 to lead to lower subjective ratings in this experiment. Power analysis revealed that, in order to have a 90 % chance of detecting differences in our data, we would need at least 10 participants (partial $\eta^2 = 0.836$, effect size 2.258, actual power 0.92). Because it is difficult to estimate a priori the correlation among repeated measures, we estimated power as though the measures were independent.

Ten participants took part in the experiment, including 4 women and 6 men. Two of them had previous experience with haptic interfaces. None of the participants reported any deficiencies in their visual or haptic perception abilities, and all of them were right-hand dominant. The experimenter explained the procedures and spent about five minutes adjusting the setup to be comfortable before the subject began the experiment.

Figure 4.9 shows the average rating versus device degrees of freedom for each tactile interaction for a representative version of the algorithm ($n = 1$, $\theta = 3°$). One observes a general trend of higher ratings with more degrees of freedom. Furthermore, the simplest interaction (*flat*) has approximately the same rating for both versions of the device controller, while more complex interactions have lower ratings that are more sensitive to controller version. All of the results, averaged among tactile interactions, are shown in Fig. 4.10. In order to compare the different algorithm and device configurations, we ran a four-way repeated-measures ANOVA. Step size during data collection, number of neighbors retrieved by function $\nu_{d,n}(\cdot)$, device degrees of freedom, and type of tactile interaction were considered as within-subject

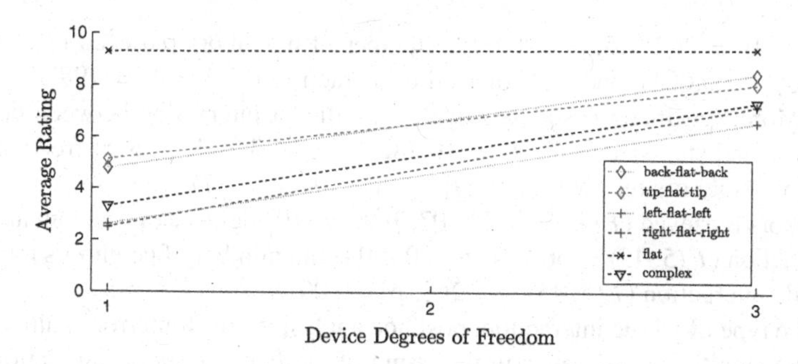

Fig. 4.9 Selected results from subjective evaluation. Degrees of freedom versus rating for each tactile interaction for a representative configuration ($n = 1$, step size angle during data collection 3°)

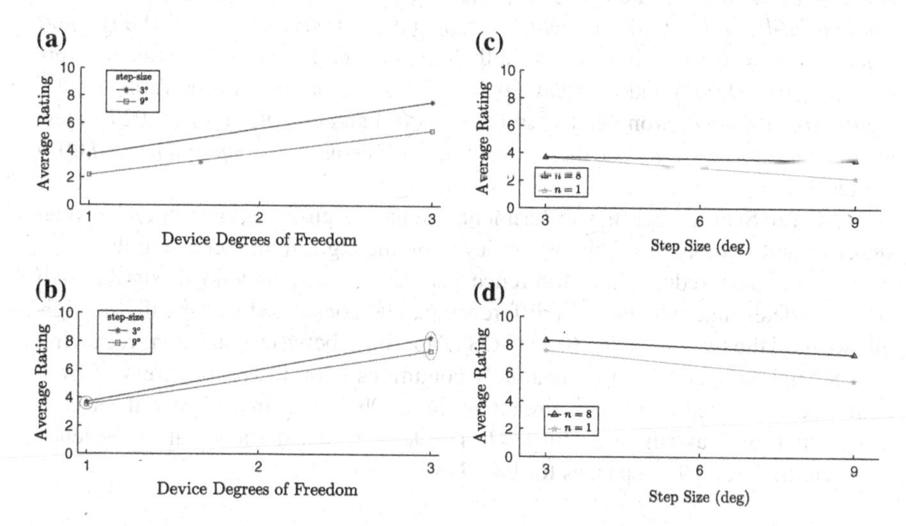

Fig. 4.10 Results of subjective evaluation averaged across tactile interactions. Degrees of freedom versus rating (**a**, **b**, **c**) and step size versus rating (**d**, **e**, **f**) are reported. Statistical analysis revealed a significant decrease of the user rating when reducing the degrees of freedom of the cutaneous device, when increasing the step size during data collection, and when reducing the number of neighbors retrieved by function $\nu_{d,n}(\cdot)$. These results agree with the findings of the objective evaluation, which are shown in Fig. 4.7. **a** Degrees of freedom versus rating, n = 1. **b** Degrees of freedom versus rating, n = 8. **c** Step size versus rating, 1-DoF configuration. **d** Step size versus rating, 3-DoF configuration

factors. In contrast to the objective analysis presented in Sect. 4.2.3.1, it is important to note that we can consider the type of tactile interaction as a within-subject factor because we had ten independent measurements for each condition, rather than a single experiment.

Ratings were subjected to the arcsine square root transformation to stabilize variance [18]. All the transformed data passed the Shapiro–Wilk normality test and Mauchly's Test of Sphericity. Sphericity was assumed for variables with only two levels of repeated measures. The test revealed a statistically significant change in the rating due to device degrees of freedom ($F(1, 9) = 2369.184$, $p < 0.001$), step size ($F(1, 9) = 202.415$, $p < 0.001$), number of neighbors retrieved ($F(1, 9) = 208.951$, $p < 0.001$), and type of tactile interaction ($F(5, 45) = 198.972$, $p < 0.001$). Moreover, there was a statistically significant interaction between degrees of freedom and step size ($F(1, 9) = 10.782$, $p = 0.009$), degrees of freedom and number of neighbors retrieved ($F(1, 9) = 6.015$, $p = 0.037$), step size and number of neighbors retrieved ($F(1, 9) = 88.307$, $p < 0.001$), degrees of freedom and tactile interaction ($F(5, 45) = 66.824$, $p < 0.001$), and number of neighbors retrieved and tactile interaction ($F(5, 45) = 5.268$, $p = 0.001$).

For the type of tactile interaction, post hoc analysis with Bonferroni adjustments revealed a significant difference in the rating of the *back-flat-back* interaction versus *left-flat-left* ($p < 0.001$), *right-flat-right* ($p < 0.001$), *flat* ($p < 0.001$), and *complex* ($p = 0.007$) interactions; in the rating of the *tip-flat-tip* interaction versus *left-flat-left* ($p < 0.001$), *right-flat-right* ($p < 0.001$), *flat* ($p < 0.001$), and *complex* ($p = 0.043$) interactions; in the rating of the *left-flat-left* interaction versus *flat* ($p < 0.001$) and *complex* ($p = 0.007$) interactions; in the rating of the *right-flat-right* interaction versus *flat* ($p < 0.001$) and *complex* ($p = 0.029$) interactions; and in the rating of the *flat* interaction versus the *complex* ($p < 0.001$) interaction.

Similar to Sect. 4.2.3.1, this experiment also has a significant interaction between step size and number of neighbors retrieved by the algorithm: increasing the number of neighbors reduces the difference between conditions with dissimilar step sizes. To determine whether this difference can be considered statistically negligible, we used the two one-sided t-test (TOST) approach between conditions retrieving $n = 8$ neighbors, considering separately conditions with different degrees of freedom. The two tested groups are circled in Fig. 4.10b. We again evaluated the acceptance criterion ε as suggested in [14]. The tests revealed statistical equivalence between the 3° and 9° step sizes for both DoF values.

4.2.4 Discussion

We ran two experiments on remote tactile interaction. The objective experiment in Sect. 4.2.3.1 evaluated the error between the tactile sensations registered by the BioTac in the remote environment and the ones actuated by the cutaneous device for six tactile interactions and 27 algorithm and device configurations. The subjective experiment in Sect. 4.2.3.2 evaluated the quality of the human user's experience for the same six interactions and eight selected algorithm and device configurations. Both experiments showed improvements (lower errors or higher subjective ratings) when increasing the degrees of freedom of the cutaneous device, decreasing the step size during data collection, and increasing the number of neighbors retrieved by the search function.

Although it is clear that increasing the degrees of freedom of the cutaneous device leads to better performance, the improvement from 2-DoF to 3-DoF is not as marked as the one from 1-DoF to 2-DoF. This difference is clear from Fig. 4.7a–c, and the statistical analysis reported in Sect. 4.2.3.1. This fact can be explained by considering that three of the six tactile interactions do not require the use of the third degree of freedom (rotation around the long axis of the finger). This dependence on tactile interaction is clear from Fig. 4.6, where the average errors of the *back-flat-back*, *tip-flat-tip*, and *flat* interactions do not change much between the conditions with two and three degrees of freedom. Moreover, from the same figure, we can also notice how the average error of the *flat* interaction does not change much across all three DoF values; this result is expected because it is a 1-DoF interaction. A similar behavior can be also spotted in Fig. 4.9, where the subjective ratings of the *flat* interaction do not differ significantly between 1-DoF and 3-DoF.

A small step size guarantees fine search spaces $\mathbb{S}_{d,*}$ and $\mathbb{M}_{d,*}$, which was found to produce higher performance in terms of both mean error and user rating (see Figs. 4.7d–f and 4.10c, d, respectively). However, as highlighted in Sect. 4.3.2, the time needed to complete the data collection phase grows cubically with the inverse of the step size. For this reason, it is important to explore solutions guaranteeing adequate performance with coarser search spaces. This work investigated the possibility of interpolating within the motor space $\mathbb{M}_{d,*}$ by retrieving multiple neighbors of the sensed point. We hypothesized that this approach would reduce the need for a small step size, providing comparable performances at larger step sizes. From equality tests in Sects. 4.2.3.1 and 4.2.3.2, we can conclude that increasing the number of neighbors retrieved does mitigate the negative effect of choosing a large step size during data collection. This finding is also clear from Figs. 4.7a–c and 4.10a, b, where lines depicting different step sizes get closer to each other when increasing the number of neighbors retrieved. However, notice also that more neighbors retrieved means a higher run-time computational load.

Although it is clear that increasing the number of neighbors retrieved leads to improved performance, it is important to notice that this relationship is not linear. As we can see from Fig. 4.7a–c, the improvement from $n = 4$ to $n = 8$ is not as large as the one from $n = 1$ to $n = 4$. These diminishing returns are due to the fact that each neighbor is weighted according to the inverse of its squared distance from the sensed point (see Eq. 4.7), and, therefore, each new neighbor counts less than all the preceding (closer) ones. It is also interesting to notice that retrieving more than one neighbor yields only a small improvement in performance, if any, for small step sizes: the 3° condition is similar across Fig. 4.7a–c and across Fig. 4.10a, b. This constant level of performance probably stems from the fact that the motor space $\mathbb{M}_{d,*}$ is already fine enough for the given tactile experience and, thus, interpolating it is not necessary.

Turning our attention back to the six tactile interactions, we recall that the *flat* interaction was rated most highly by subjects, as shown by Fig. 4.9 and the analysis of Sect. 4.2.3.2. Because this interaction matched the motion available from the 1-DoF device, it was rendered very well even when driving the three servo motors together (left side of Fig. 4.9). The performance of the other five interactions dramatically improved when the device used all three of its degrees of freedom (right side of Fig. 4.9). The post-hoc analysis of Sect. 4.2.3.2 showed that the *left-flat-left* and *right-flat-right* interactions performed significantly worse than the other tactile interactions. Four of the subjects also stated this opinion in their post-experiment questionnaire.

We believe the poor performance of *left-flat-left* and *right-flat-right* was caused by two main factors. The first reason is related to the distribution of the impedance-sensing electrodes on the surface of the BioTac core. Each side of the sensor includes only three electrodes, and we used Euclidean distance when looking for neighbors; consequently, the tactile sensations caused by platform contact on the sides of the BioTac may be less noticeable than tactile information from other types of interactions. The second factor centers on the shape of the platform and the positioning of the servo motors. During the data collection phase, we noticed that the seclected cutaneous device struggles to apply pressure to only the lateral electrodes: the motor at the front of the device stimulates *all* the distal electrodes, while the two motors placed at the rear leave the distal lateral part of the BioTac untouched. If such interactions were important to the chosen task, one could design a tactile sensor and/or cutaneous device that were more adept at measuring and applying lateral contacts.

4.3 Cutaneous Feedback System for Remote Interaction: Contact Deformations and Vibrotactile Cues

Section 4.2 presented and evaluated the tactile rendering algorithm considering only contact deformations. This section extends the approach to include also vibrotactile cues and evaluates the system in a challenging robot-assisted surgery scenario. Similarly to what we have already seen in Sect. 4.2, the system is composed of a BioTac tactile sensor and a cutaneous display device. However, this time, the cutaneous device is also equipped with a vibrotactile motor. Contact deformations and vibrations sensed by the BioTac are directly mapped to input commands for the cutaneous device's motors using an extended version of the model-free data-driven algorithm described in Sect. 4.2.2.

We tested the proposed approach in a palpation task using the da Vinci Surgical System. The experimental setup is shown in Fig. 4.11. Eighteen subjects were asked to detect the orientation of a plastic stick hidden in a tissue phantom hearth model. Results show that providing cutaneous feedback significantly improved the task performance in terms of completion time, pressure exerted on the tissue model, and absolute error in detecting the orientation of the plastic stick.

4.3.1 Sensing and Actuation Systems

As detailed in Sect. 4.2.1, the BioTac tactile sensor is capable of sensing both contact deformations and vibrations. Contact forces deform the elastic skin and the underlying conductive fluid, changing the impedances of 19 electrodes distributed over

Fig. 4.11 System setup. The haptic system is composed of a BioTac tactile sensor, in charge of registering contact forces and vibrations at the operating table, and a custom cutaneous device, in charge of applying those forces and vibrations to the surgeon. The BioTac is attached to a da Vinci tool, and the cutaneous device is attached to the robot's corresponding master controller. The BioTac follows the motions of the operator's finger

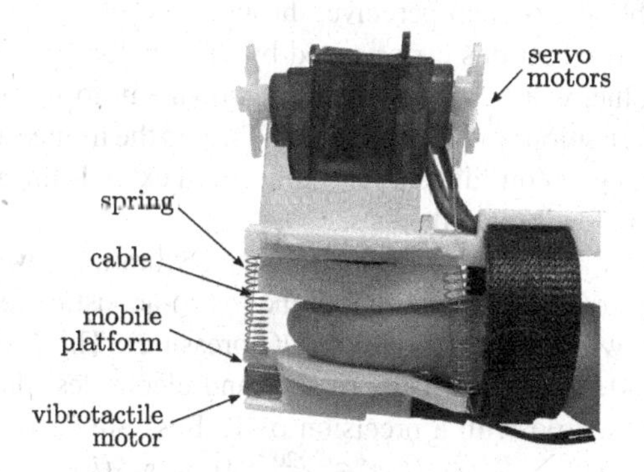

Fig. 4.12 The cutaneous device frame houses three servo motors, and a mobile platform holding one vibrotactile motor is suspended beneath the operator's fingertip by three cables. By controlling the cable lengths, the servos move the mobile platform in three-dimensional space to apply contact deformations to the fingertip, while the vibrotactile motor conveys fingertip vibrations. This device is an augmented version of the one presented in Sect. 4.2

the surface of the rigid core. Vibrations in the skin propagate through the fluid and are detected as AC signals by a hydro-acoustic pressure sensor, while this sensor's DC value indicates the steady-state pressure of the fluid. As shown in Fig. 4.11, we attached the BioTac sensor to the da Vinci slave robotic tool through a custom plastic nail.

The haptic tactile interface employed is shown in Fig. 4.12. It is an augmented version of the one presented in Sect. 4.2. The device frame houses three servo motors and is rigidly attached to the grip interface of the da Vinci master controller as shown in Fig. 4.11. The device fastens to the user's index finger with a strap between the PIP and DIP joints. A mobile platform holding one vibrotactile motor is suspended beneath the operator's fingertip by three cables; compression springs around the cables hold the mobile platform in a reference configuration away from the fingertip. By controlling the cable lengths, the servos orient and translate the mobile platform in three-dimensional space to apply planar deformations to the fingertip, while the vibrotactile motor conveys fingertip vibrations. As in Sect. 4.2, the servo motors used are Sub-Micro Servo 3.7 g motors, which are position controlled and can exert up to 39 N·mm torque. Our vibrotactile motor is a Forcereactor short-vibration feedback motor (Alps Electric, Japan). A brief video showing the device is available as supplemental material at http://extras.springer.com/978-3-319-25455-5 and at http://goo.gl/u9CXSw.

4.3.2 Mapping Between Remote Sensed Data and Motor Commands

Our goal is to enable the user to perceive, through the fingertip cutaneous device, the deformations and vibrations experienced by the BioTac as it interacts with the patient's tissue. In other words, we aim to find a good many-to-few mapping between the rich sensory information provided by the BioTac and the limited actuation capabilities of the fingertip cutaneous device. This algorithm extends the approach detailed in Sect. 4.2.2 to include vibrotactile cues.

Since our focus is in sensing deformations and vibrations, we consider the 19 electrode impedance readings, the output of the hydro-acoustic pressure sensor (AC pressure), and its low-pass filtered output (DC pressure). The BioTac registers the AC pressure at 2200 Hz, and the DC pressure and electrodes' data at 100 Hz. All these quantities are sensed with a precision of 12 bits. Similarly to Sect. 4.2.2, let $\boldsymbol{s_d(k)} \in \mathbb{S}_d = \{(s_{d,1}(k), \ldots, s_{d,20}(k)) \in \mathbb{Z}^{20} : 0 \leq s_{d,i}(k) \leq 4095\}$ be a vector containing the electrodes and DC pressure values sensed at instant k, and $s_v(k) \in \mathbb{Z} \in \mathbb{S}_v = \{s_v(k) \in \mathbb{Z} : 0 \leq s_d(k) \leq 4095\}$ the AC pressure sensed at instant k. In contrast, our cutaneous device uses three position-controlled motors and one vibrotactile motor. Let $\boldsymbol{m_d(k)} \in \mathbb{M}_d = \{(m_{d,1}(k), m_{d,2}(k), m_{d,3}(k)) \in \mathbb{R}^3 : 30° \leq m_{d,i}(k) < 200°\}$ be a vector containing the commanded angles for the three position-controlled motors at instant k, and $m_v(k) \in \mathbb{M}_v = \{m_v(k) \in \mathbb{R} : 0 \leq m_v(k) \leq 1\}$ the commanded signal for the vibrotactile motor at instant k. Note that we are again neglecting quantization in the motor position outputs, and, in order to simplify the notation further, the sampling time index k will be omitted.

Our objective is thus to map a given BioTac sensation, defined by $\boldsymbol{s_d}$ and s_v, to a congruent configuration of the mobile platform $\boldsymbol{m_d}$ and signal for the vibrotactile motor m_v. At first, in order to understand how the motion of the mobile platform affects the tactile sensor and the way it senses vibrations, we placed the BioTac inside the cutaneous device and tested the effect of different platform configurations as described in Sect. 4.3.2.1. Using these data, contact deformations sensed by the BioTac are mapped to input commands for the cutaneous device's servo motors as described in Sect. 4.3.2.2 (similarly to what we have already done in Sect. 4.2.2). Finally, vibrations sensed by the BioTac are mapped to input commands for the cutaneous' device vibrotactile motor as described in Sect. 4.3.2.3. The algorithm is summarized in Fig. 4.13.

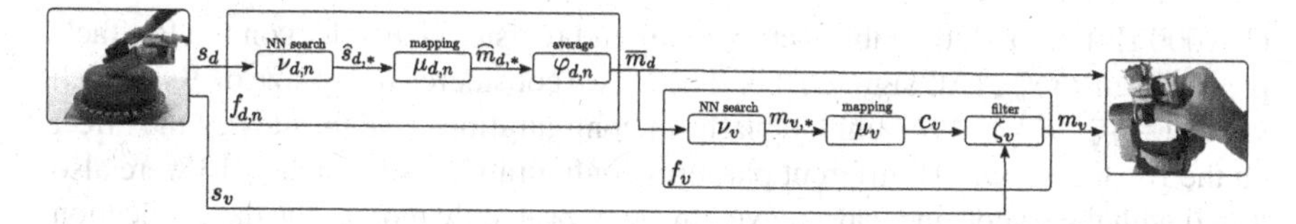

Fig. 4.13 Rendering algorithm for contact deformations and vibrations. The BioTac registers a tactile sensation $[\boldsymbol{s}_d \ \ s_v]^T$ at the remote environment. Function $\nu_{d,n}(\cdot)$ looks for the n closest points contained in $\mathbb{S}_{d,*}$, recorded during data collection. These points are then mapped by $\mu_{d,n}(\cdot)$ to their corresponding motor angle triplets. Function $\varphi_{d,n}(\cdot)$ then averages those points to find the angle triplet $\overline{\boldsymbol{m}}_d$ to be actuated by the device. Once the platform configuration is defined, function $\nu_v(\cdot)$ looks for the closest platform configuration contained in $\mathbb{M}_{v,*}$, recorded during data collection, and retrieves the corresponding transfer function. Finally, $\zeta_v(\cdot)$ filters the vibrations sensed by the BioTac accordingly. In our experimental evaluation we considered $n = 8$ and $|\mathbb{S}_{d,*}| = 39{,}304$, since Sect. 4.2 has found these values to be a good trade-off between performance and computational load

4.3.2.1 Data Collection

As shown in Fig. 4.14, the BioTac was placed between the foam and the mobile platform, in the same way a human user would wear the device (see Fig. 4.12). We then moved the mobile platform to a wide range of configurations and registered the effect of each of these configurations on the BioTac, saving both the commanded servo motor angles $\boldsymbol{m}_{d,*}$ and the resulting effect on the tactile sensor's electrodes and DC pressure $\boldsymbol{s}_{d,*}$. Using a moderate step size of 5° yields $\left(\frac{200°-30°}{5°}\right)^3 = 39{,}304$ unique platform configurations. The platform was held in each configuration for 0.2 s, and the values gathered by the BioTac were arithmetically averaged. During this process, in order to understand how different platform configurations affect the way the BioTac feels vibrations, we also played a two-second-long swept sine wave

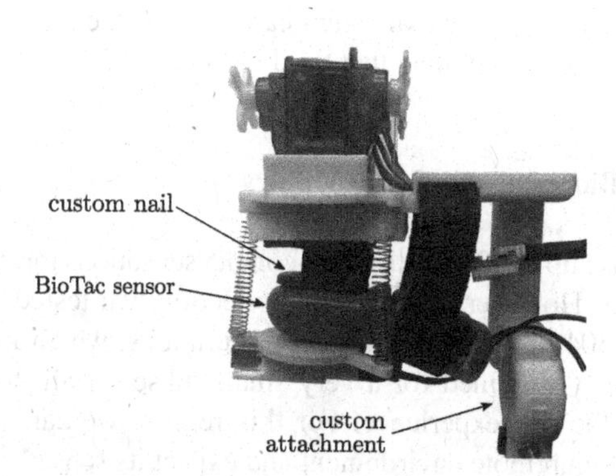

Fig. 4.14 Data collection. The BioTac was placed inside the cutaneous device, and the platform was moved to 39,304 different configurations. Servo motor inputs $\boldsymbol{m}_{d,*}$ and the resulting effect on the tactile sensor's electrodes and DC pressure $\boldsymbol{s}_{d,*}$ were recorded. For 4913 of these configurations, we also played a two-second-long swept sine wave (1–1000 Hz) through the vibrotactile motor and registered its effect on the BioTac's pressure sensor $s_{v,*}$ (AC signals)

(1–1000 Hz) through the vibrotactile motor and registered its effect on the BioTac's pressure sensor $s_{v,*}$ (AC signals). For this test we considered a step size of 10°, which yields $(\frac{200°-30°}{10°})^3 = 4913$ unique platform configurations. We thus tested the effect on the BioTac of 39,304 different platform configurations, of which 4913 were also tested with the swept sine wave played by the vibrotactile motor. The data collection lasted approximately 22 h.

At the end of data collection, we were able to evaluate the mapping function

$$\begin{aligned} \mu_d &: \mathbb{S}_{d,*} \rightarrow \mathbb{M}_{d,*}, \\ \mu_d(\boldsymbol{s_{d,*}}) &= \boldsymbol{m_{d,*}}, \end{aligned} \tag{4.9}$$

which links the BioTac sensed data (electrodes and DC pressure) to the corresponding tested servo motor inputs (similarly to Sect. 4.2.2). Set $\mathbb{M}_{d,*} \subset \mathbb{M}_d$ contains all the angle triplets actuated during data collection, and $\mathbb{S}_{d,*} \subset \mathbb{S}_d$ contains all the resulting sensed values registered by the BioTac. Moreover, for the 4913 platform configurations tested with the swept sine wave, we were also able to estimate the transfer function between the vibrations sensed by the BioTac and the swept sine wave played through the vibrotactile motor. We used MATLAB's function `tfest`. Each transfer function contained 6 poles and 6 zeros. We were then able to also evaluate the mapping function

$$\begin{aligned} \mu_v &: \mathbb{M}_{v,*} \rightarrow \mathbb{R}^{14}, \\ \mu_v(\boldsymbol{m_{v,*}}) &= \boldsymbol{c_v}, \end{aligned} \tag{4.10}$$

which links the 4913 tested platform configurations $\boldsymbol{m_{v,*}}$ to the corresponding transfer function's coefficients $\boldsymbol{c_v}$. Set $\mathbb{M}_{v,*} \subset \mathbb{M}_{d,*}$ contains the angle triplets tested with the swept sine wave during data collection.

Function $\mu_d(\cdot)$ presented here is similar to the same function presented in Eq. (4.1). However, sets $\mathbb{S}_{d,*}$ and $\mathbb{M}_{d,*}$ have different cardinalities, i.e., the step size used for data collection here is different than that used in Sect. 4.2.2.

4.3.2.2 From the BioTac to the Servo Motors

The BioTac can sense up to 4096^{20} different contact sensations through its electrodes and pressure sensor. However, during data collection, we tested the effect on the BioTac of only 39,304 different platform configurations, which inevitably led to a mapping function $\mu_d(\cdot)$ defined for a very small subset of all the possible tactile sensations the BioTac can experience. For this reason, we cannot simply deploy the sensor in a random remote environment and expect its sensed points to be in the domain of our mapping function. We thus need a function that maps a *generic* sensed point $\boldsymbol{s_d} \in \mathbb{S}_d$ to one in our mapping function's domain $\mathbb{S}_{d,*}$.

As extensively discussed in Sect. 4.2, we address this problem by looking for the n-points in our domain closest to the sensed one, and thus define

$$\nu_{d,n} : \mathbb{S}_d \rightarrow \mathbb{S}^n_{d,*},$$
$$\nu_{d,n}(\boldsymbol{s}_d) = \begin{bmatrix} \boldsymbol{s}_{d,*,1} \\ \boldsymbol{s}_{d,*,2} \\ \vdots \\ \boldsymbol{s}_{d,*,n} \end{bmatrix} = \widehat{\boldsymbol{s}}_{d,*}, \tag{4.11}$$

as the function that maps a generic point $\boldsymbol{s}_d \in \mathbb{S}_d$, sensed by the BioTac, to the n closest one in $\mathbb{S}_{d,*}$. We implemented again the nearest point search using the Approximate Nearest Neighbour (ANN) C++ library by Mount and Arya [12], using the 20-dimensional Euclidean distance metric and $n = 8$. As discussed in Sect. 4.2.4, retrieving the eight closest points is a good trade-off between performance and computational load.

In order to evenly weight the twenty elements of the sensed data when computing the distance, we divided each component of $\boldsymbol{s}_d$ and $\boldsymbol{s}_{d,*}$ by the corresponding standard deviation observed during data collection, so that the standard deviation of each component of the vectors in $\mathbb{S}_{d,*}$ becomes 1. The same operation was applied at run time to $\boldsymbol{s}_d$

We can now map the n points retrieved by $\nu_{d,n}(\cdot)$ to their corresponding motor angle triplets registered during data collection,

$$\mu_{d,n} : \mathbb{S}^n_{d,*} \rightarrow \mathbb{M}^n_{d,*},$$
$$\mu_{d,n}(\widehat{\boldsymbol{s}}_{d,*}) = \mu_{d,n}\left(\begin{bmatrix} \boldsymbol{s}_{d,*,1} \\ \boldsymbol{s}_{d,*,2} \\ \vdots \\ \boldsymbol{s}_{d,*,n} \end{bmatrix}\right) = \begin{bmatrix} \mu_d(\boldsymbol{s}_{d,*,1}) \\ \mu_d(\boldsymbol{s}_{d,*,2}) \\ \vdots \\ \mu_d(\boldsymbol{s}_{d,*,n}) \end{bmatrix} \tag{4.12}$$
$$= \begin{bmatrix} \boldsymbol{m}_{d,*,1} \\ \boldsymbol{m}_{d,*,2} \\ \vdots \\ \boldsymbol{m}_{d,*,n} \end{bmatrix} = \widehat{\boldsymbol{m}}_{d,*},$$

Finally, we can average $\widehat{\boldsymbol{m}}_{d,*} \in \mathbb{M}^n_{d,*}$ to a single angle triplet for the servo motors as

$$\varphi_{d,n} : \mathbb{M}^n_{d,*} \rightarrow \mathbb{M}_d,$$
$$\varphi_{d,n}(\widehat{\boldsymbol{m}}_{d,*}) = \varphi_d\left(\begin{bmatrix} \boldsymbol{m}_{d,*,1} \\ \boldsymbol{m}_{d,*,2} \\ \vdots \\ \boldsymbol{m}_{d,*,n} \end{bmatrix}\right) = \overline{\boldsymbol{m}}_d, \tag{4.13}$$

considering a simple inverse squared distance mean that weights angle triplets according to the inverse squared distance between the corresponding point in $\mathbb{S}_{d,*}$ and the one sensed by the BioTac (as in Sect. 4.2). Vector $\overline{\boldsymbol{m}}_d \in \mathbb{M}_d$ is our final input for the servo motors.

For the sake of simplicity, we can also define

$$\begin{aligned} f_{d,n} &: \mathbb{S}_d \rightarrow \mathbb{M}_d, \\ f_{d,n}(\boldsymbol{s_d}) &= \varphi_d(\mu_{d,n}(\nu_d(\boldsymbol{s_d}))) = \overline{\boldsymbol{m}}_{\boldsymbol{d}}, \end{aligned} \tag{4.14}$$

as the function that maps a generic point sensed by the BioTac to an angle triplet for the servo motors.

Function $f_{d,n}(\cdot)$ presented here is similar to the same function presented in Eq. (4.8). However, here we took into account only the case with $n = 8$, considering a step size during data collection of 5°. Moreover, as detailed in the next section, $\overline{\boldsymbol{m}}_{\boldsymbol{d}}$ will be here also used to map the vibrations sensed by the BioTac to suitable input commands for the vibrotactile motor.

4.3.2.3 From the BioTac to the Vibrotactile Motor

During data collection we also tested how 4913 different platform configurations affected the way the BioTac senses vibrations. However, our designated input for the servo motor $\overline{\boldsymbol{m}}_{\boldsymbol{d}}$ is defined in all $\mathbb{M}_d$, which is much larger than our data collection set $\mathbb{M}_{v,*}$. Similarly to what we have done in Sect. 4.3.2.2, we address this problem by looking for the platform configuration in $\mathbb{M}_{v,*}$ closest to $\overline{\boldsymbol{m}}_{\boldsymbol{d}}$, and thus define

$$\begin{aligned} \nu_v &: \mathbb{M}_d \rightarrow \mathbb{M}_{v,*}, \\ \nu_v(\overline{\boldsymbol{m}}_{\boldsymbol{d}}) &= \boldsymbol{m}_{\boldsymbol{v},*}, \end{aligned} \tag{4.15}$$

as the function that maps a generic platform configuration $\overline{\boldsymbol{m}}_{\boldsymbol{d}} \in \mathbb{M}_d$ to the closest one tested with the swept sine wave during data collection.

We can now retrieve the corresponding transfer function's coefficients, as defined in Eq. (4.10), and filter the vibrations sensed by the BioTac in the operative environment accordingly,

$$\begin{aligned} \zeta_v &: (\mathbb{S}_v, \mathbb{R}^{14}) \rightarrow \mathbb{M}_v, \\ \zeta_v(s_v, \boldsymbol{c_v}) &= \zeta_v(s_v, \mu_v(\boldsymbol{m}_{\boldsymbol{v},*})) = m_v. \end{aligned} \tag{4.16}$$

We filtered the signal using a floating-point implementation of a IIR filter [19]. Value m_v is our final input for the vibrotactile motor.

For the sake of simplicity, we can also define

$$\begin{aligned} f_v &: (\mathbb{S}_v, \mathbb{M}_d) \rightarrow \mathbb{M}_v, \\ f_v(s_v, \overline{\boldsymbol{m}}_{\boldsymbol{d}}) &= \zeta_v(s_v, \mu_v(\nu_v(\overline{\boldsymbol{m}}_{\boldsymbol{d}}))) = m_v, \end{aligned} \tag{4.17}$$

as the function that filters a vibration sensed by the BioTac to an input for our vibrotactile motor, taking into account how the current platform configuration affects the way vibrations are propagated from the motor to the sensor.

4.3.3 Experimental Evaluation

We evaluated the proposed haptic system by carrying out a palpation task using a da Vinci Surgical System. The experimental setup is shown in Fig. 4.11. The cutaneous device is attached to the robot's right master controller, and the BioTac to the corresponding da Vinci tool, so that it follows the motion of the operator's finger. The remote environment is composed of a tissue phantom heart model made from Ecoflex 0010 (Smooth-On Inc, PA, USA) and brown dye. As shown in Fig. 4.15, a plastic stick is embedded into the tissue model at 1.5 mm from the surface, and it is not visible from the outside. The plastic stick simulates the presence of a calcified artery [20].

Eighteen participants took part in the experiment, including 7 women and 11 men. Five of them had previous experience with haptic interfaces. None of the participants reported any deficiencies in their visual or haptic perception abilities, and all of them were right-hand dominant. The experimenter explained the procedures and spent about three minutes adjusting the setup to be comfortable before the subject began the experiment. Moreover, the subjects had about two minutes to practice the control of the slave tools through the master console.

The task consisted of exploring the tissue model to try to detect the orientation of the hidden plastic stick. As shown in Fig. 4.11, the model was placed on a rounded plastic base with 36 ticks. The ticks indicate the 18 possible orientations of the hidden stick, reporting angles from 0° to 180°, with a step size of 10°. Subjects were asked to explore the tissue model and tell the experimenter the orientation of the stick. The task started when the BioTac touched the tissue model for the very first time and ended when the subject told the experimenter the estimated orientation. Each participant made twelve trials of the palpation task, with four repetitions for each feedback condition proposed:

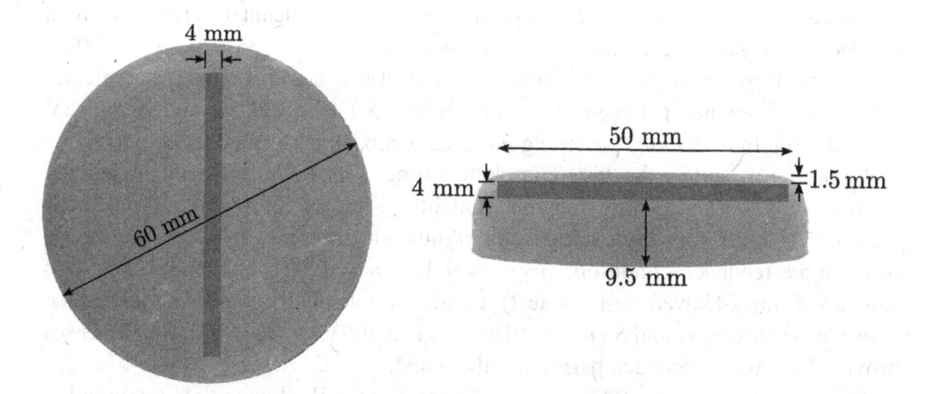

Fig. 4.15 Tissue phantom hearth model. The diameter and thickness of the phantom tissue heart model are 60 and 15 mm, respectively. A plastic stick with diameter of 4 mm, which mimics a calcified artery, is embedded at the depth of 1.5 mm from the surface

- no force feedback (condition N),
- cutaneous feedback provided by the servo motors (condition S),
- cutaneous feedback provided by the servo and vibrotactile motors (condition SV).

In condition N, the servo and vibrotactile motors were not active and the mobile platform was always in contact with the subject's fingertip. In condition S, the servo motors moved the mobile platform as described in Sect. 4.3.2.2 and the vibrotactile motor was not active. In condition SV, the servo motors moved the mobile platform as in condition S and the vibrotactile motor provided vibrations to the subjects as described in Sect. 4.3.2.3. Subjects were always able to see the operative environment through the standard stereoscopic monitor of the da Vinci System. The order we provided the feedback conditions was randomized to test all six possible combinations exactly three times. At the end of each condition, subjects were asked to rate, on two sliders going from 0 to 10, "how easy was it to detect the orientation of the plastic stick?" and "how confident were you in detecting the orientation of the plastic stick?". A score of 0 meant "very difficult" ("not at all confident") while a score of 10 meant "very easy" ("very confident"). Moreover, at the end of the experiment, subjects were asked to choose which feedback condition was the most and least effective at letting them detect the orientation of the plastic stick. A video of the experiment is available as supplemental material at http://extras.springer.com/978-3-319-25455-5 and at http://goo.gl/dSzdkM.

In order to evaluate the performance of the considered feedback conditions, we evaluated (1) the absolute error in detecting the orientation of the plastic stick, (2) the task completion time, and (3) the root mean square (RMS) pressure exerted by the BioTac on the tissue model.

Figure 4.16a shows the absolute error in detecting the orientation of the hidden stick for the three experimental conditions. The collected data passed Shapiro-Wilk normality test. Mauchly's Test of Sphericity indicated that the assumption of sphericity had been violated ($\chi^2(2) = 6.245$, $p = 0.044$). A repeated-measure ANOVA with a Greenhouse-Geisser correction showed a statistically significant difference between the means of the three feedback conditions ($F_{1.512,25.696} = 33.890$, $p < 0.001$, a $= 0.05$). Post-hoc analysis (Games-Howell post-hoc test) revealed a statistically significant difference between conditions N and S ($p < 0.001$), and N and SV ($p < 0.001$). In both cases providing no force feedback performed the worst.

Figure 4.16b shows the task completion time. The collected data passed the Shapiro-Wilk normality test and the Mauchly's Test of Sphericity. A repeated-measure ANOVA showed a statistically significant difference between the means of the three feedback conditions ($F_{2,34} = 9.342$, $p = 0.001$, a $= 0.05$). Post-hoc analysis (Games-Howell post-hoc test) revealed a statistically significant difference between conditions N and S ($p = 0.019$), and N and SV ($p = 0.002$). In both cases providing no force feedback performed the worst.

Figure 4.16c shows the RMS pressure exerted by the BioTac on the tissue model, registered through its hydro-acoustic pressure sensor (DC data). The collected data

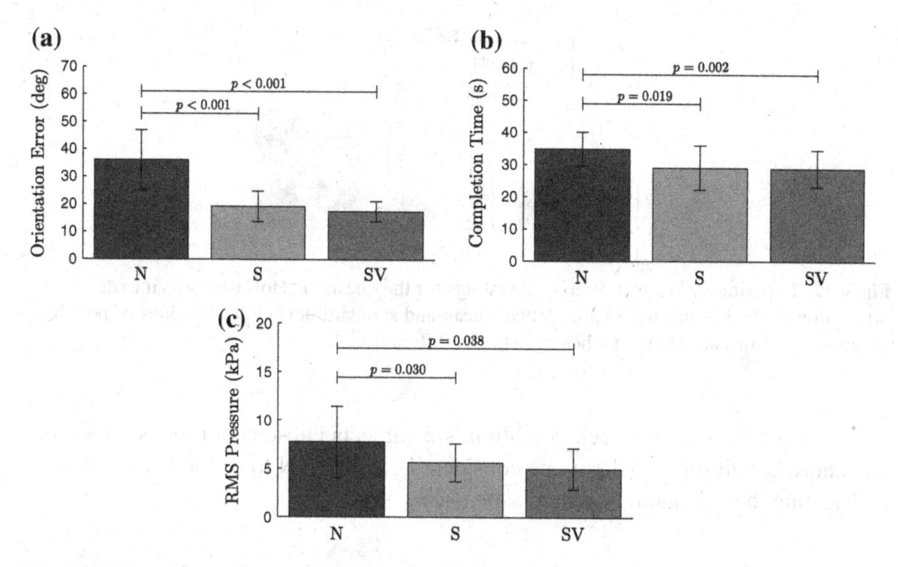

Fig. 4.16 Experimental results. Absolute orientation error, completion time, and RMS pressure for conditions providing no force feedback (N), cutaneous feedback by the servo motors (S), and cutaneous feedback by the servo and vibrotactile motors (SV) are plotted (mean and standard deviation). Lower values of these metrics indicate higher performances in completing the palpation task. P-values of post-hoc group comparisons are reported when statistically different. **a** Absolute orientation error. **b** Completion time. **c** RMS pressure

passed the Shapiro-Wilk normality test. Mauchly's Test of Sphericity indicated that the assumption of sphericity had been violated ($\chi^2(2) = 9.120$, $p = 0.010$). A repeated-measure ANOVA with a Greenhouse-Geisser correction showed a statistically significant difference between the means of the three feedback conditions ($F_{1.394,23.702} = 6.908$, $p = 0.009$, a $= 0.05$). Post-hoc analysis (Games-Howell post-hoc test) revealed a statistically significant difference between conditions N and S ($p = 0.030$), and N and SV ($p = 0.038$). In both cases providing no force feedback performed the worst.

Finally, we analyzed the ratings given by the subjects at the end of each feedback condition. All the data passed the Shapiro-Wilk normality test and the Mauchly's Test of Sphericity. A repeated-measure ANOVA showed a statistically significant difference between the means of the three feedback conditions for both questions ("how easy", $F_{2,34} = 52.460$, $p < 0.001$, a $= 0.05$; "how confident", $F_{2,34} = 50.517$, $p < 0.001$, a $= 0.05$). Post-hoc analysis (Games-Howell post-hoc test) revealed statistically significant difference between conditions N and S ($p < 0.001$), and N and SV ($p < 0.001$) for both questions. Figure 4.17 shows the mean ratings for the three feedback conditions in the first question. The second one showed similar results (N: 3.45, S: 7.13, SV: 7.71). Conditions providing cutaneous feedback were also preferred in the post-experimental questionnaire. Eleven subjects chose condition SV

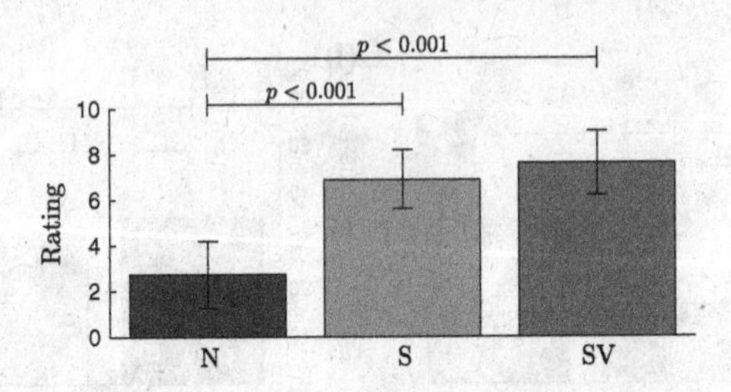

Fig. 4.17 Experimental results. Preference ratings for the question "How easy was it to detect the orientation of the plastic stick?" are plotted (mean and standard deviation). P-values of post-hoc group comparisons are reported when statistically different

as the most effective feedback condition, six subjects chose condition S, and only one chose condition N. All but one considered condition N to be the least effective at detecting the orientation of the plastic stick.

4.3.4 Discussion

Adding cutaneous feedback significantly improved the task performance in all the considered metrics. Moreover, conditions providing cutaneous feedback were highly preferred by the subjects. However, we did not find any significant difference between the two conditions providing cutaneous feedback, S and SV. Adding vibrations did not thus seem to improve the performance of the given task.

Our hypothesis is that this result is mainly due to the way subjects explored the tissue model. We noticed in fact two different strategies being used. Referring to the classification of hand exploratory movements described by Lederman and Klatzky [21], we can describe the first strategy as a "pressure" exploratory movement and the second one as a combination of the "pressure" and the "lateral motion" exploratory movements. We will refer to them as "pressure strategy" and "dragging strategy", respectively. A video showing these two exploratory strategies is available as supplemental material at http://extras.springer.com/978-3-319-25455-5 and at http://goo.gl/m2Oppn. Eleven subjects used the "pressure strategy" while seven used the "dragging strategy". In order to understand how different strategies affected the performance of the palpation task, we performed a statistical analysis considering only the seven subjects who used the "dragging strategy". All the collected data passed the Shapiro-Wilk normality test. Mauchly's Test of Sphericity indicated that the assumption of sphericity had been violated in the orientation error data ($\chi^2(2) = 7.608$, $p = 0.022$). A repeated-measure ANOVA showed a statistically significant difference between the means of the three feedback modalities in all the metrics considered (orientation error, $F_{1.123,6.735} = 18.891$, $p = 0.003$, $a = 0.05$; completion time, $F_{2,12} = 8.291$, $p = 0.005$, $a = 0.05$; RMS pressure, $F_{2,12} = 11.935$, $p = 0.001$, $a = 0.05$). Post-hoc analysis (Games-Howell post-hoc

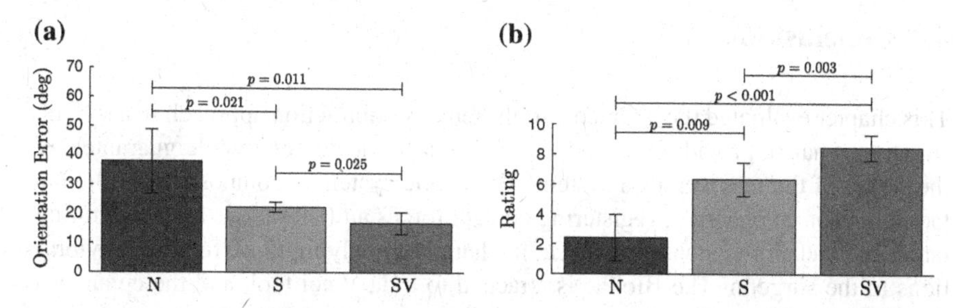

Fig. 4.18 Experimental result for the "dragging strategy". Absolute orientation error and preference ratings for the question "How easy was it to detect the orientation of the plastic stick?" are plotted (mean and standard deviation), considering only the data from the seven subjects who used the "dragging strategy". P-values of post-hoc group comparisons are reported when statistically different. **a** Absolute orientation error. **b** Preference ratings

test) revealed statistically significant difference between conditions N and S (orientation error, $p = 0.021$; completion time, $p = 0.013$; RMS pressure, $p = 0.015$), and N and SV (orientation error, $p = 0.011$; completion time, $p = 0.017$; RMS pressure, $p = 0.023$) in all conditions. Conditions S and SV were found statistically different in the orientation error metric ($p = 0.025$). Finally, we analyzed the ratings given by the seven subjects at the end of each feedback condition. All the data passed the Shapiro-Wilk normality test and the Mauchly's Test of Sphericity. A repeated-measure ANOVA showed a statistically significant difference between the means of the three feedback conditions for both questions ("how easy", $F_{2,12} = 46.790$, $p < 0.001$, a $= 0.05$; "how confident", $F_{2,12} = 37.266$, $p < 0.001$, a $= 0.05$). Post-hoc analysis (Games-Howell post-hoc test) revealed statistically significant difference between conditions N and S ("how easy", $p = 0.009$; "how confident", $p = 0.008$), and N and SV ($p < 0.001$) for both questions. Conditions S and SV were found statistically different in one of the questions ("how easy", $p = 0.009$). All the seven subjects who used the "dragging strategy" found condition SV to be the most effective at letting them detect the orientation of the plastic stick. Figure 4.18 shows the absolute orientation error and the mean preference ratings for the three feedback conditions considering only the seven subjects who used the "dragging strategy".

For these seven subjects the vibrotactile feedback provided by our device was thus useful to detect the orientation of the hidden plastic stick. This results can be explained by considering that, while dragging the BioTac on the tissue surface, the vibrotactile feedback may have helped the detection of the transient due to the presence of the hidden stick. On the other hand, in the "pressure strategy", vibrations may have helped the rendering of the sensation of breaking and making contact with the tissue surface, but they have probably provided little information about the orientation of the hidden stick.

4.4 Conclusions

This chapter evaluated the cutaneous-only sensory subtraction approach in a robotic surgery scenario, providing cutaneous feedback to the surgeon while guaranteeing the safety of the teleoperation system. The haptic system is composed of a BioTac tactile sensor, in charge of registering contact forces and vibrations at the operating table, and a custom cutaneous device, in charge of applying those forces and vibrations to the surgeon. The BioTac is attached to a da Vinci tool, and the cutaneous device is attached to the robot's corresponding master controller. Contact deformations and vibrations sensed by the BioTac are directly mapped to input commands for the cutaneous device's motors using a novel model-free data-driven algorithm. The algorithm is compatible with any fingertip mechanical sensor and mechanical actuation system and does not use any kind of skin deformation model.

First, we evaluated the algorithm considering only contact deformations. One experiment evaluated the error between the tactile sensations measured by the BioTac in a remote environment and the ones actuated by the cutaneous device for six remote tactile interactions and 27 algorithm and device configurations. The average error in the best condition was 3.0 % of the full 12-bit scale the BioTac can reach. Another experiment evaluated the subjective experience of ten users for the same six interactions and eight selected algorithm and device configurations. The average rating for the best condition was 8.2 out of 10.

Finally, we extended the approach to include also vibrotactile cues. We tested this approach in a palpation task using the da Vinci Surgical System. Eighteen subjects were asked to detect the orientation of a plastic stick hidden in a tissue phantom hearth model. Results show that providing cutaneous feedback significantly improved the task performance in terms of absolute error in detecting the orientation of the plastic stick, completion time, and pressure exerted on the tissue phantom. Moreover, subjects highly preferred conditions providing cutaneous feedback over the one not providing any kind of force feedback.

References

1. D. Prattichizzo, C. Pacchierotti, G. Rosati, Cutaneous force feedback as a sensory subtraction technique in haptics. IEEE Trans. Haptics **5**(4), 289–300 (2012)
2. C. Pacchierotti, F. Chinello, M. Malvezzi, L. Meli, D. Prattichizzo, Two finger grasping simulation with cutaneous and kinesthetic force feedback, in *Haptics: Perception, Devices, Mobility, and Communication* (2012), pp. 373–382
3. C. Pacchierotti, A. Tirmizi, D. Prattichizzo, Improving transparency in teleoperation by means of cutaneous tactile force feedback. ACM Trans. Appl. Percept. **11**(1), 4:1–4:16 (2014)
4. C. Pacchierotti, F. Chinello, D. Prattichizzo, Cutaneous device for teleoperated needle insertion, in *Proceeding of the 4th IEEE RAS EMBS International Conference on Biomedical Robotics and Biomechatronics (BioRob)* (2012), pp. 32–37
5. C. Pacchierotti, A. Tirmizi, G. Bianchini, D. Prattichizzo, Enhancing the performance of passive teleoperation systems via cutaneous feedback. IEEE Trans. Haptics (2015) (In Press)

6. C. Pacchierotti, D. Prattichizzo, K.J. Kuchenbecker, Cutaneous feedback of fingertip deformation and vibration for palpation in robotic surgery. IEEE Trans. Biomed. Eng. (2015) (In Press)
7. C. Pacchierotti, D. Prattichizzo, K.J. Kuchenbecker, Displaying sensed tactile cues with a fingertip haptic device. IEEE Trans. Haptics (2015) (In Press)
8. N. Wettels, G.E. Loeb, Haptic feature extraction from a biomimetic tactile sensor: force, contact location and curvature, in *Proceedings of the IEEE International Conference on Robotics and Biomimetics* (2011), pp. 2471–2478
9. J.A. Fishel, G.E. Loeb, Sensing tactile microvibrations with the biotac-comparison with human sensitivity, in *Proceedings of the IEEE RAS & EMBS International Conference on Biomedical Robotics and Biomechatronics (BioRob)* (2012), pp. 1122–1127
10. C. Pacchierotti, L. Meli, F. Chinello, M. Malvezzi, D. Prattichizzo, Cutaneous haptic feedback to ensure the stability of robotic teleoperation systems. Int. J. Robot. Res. (2015) http://ijr.sagepub.com/content/early/2015/10/15/0278364915603135.abstract (in press)
11. G. Rosati, F. Oscari, C. Pacchierotti, D. Prattichizzo, Effects of kinesthetic and cutaneous stimulation during the learning of a viscous force field. IEEE Trans. Haptics Spec. Issue Haptics Rehabil. Neural Eng. **7**(2), 251–263 (2014)
12. D.M. Mount, S. Arya, ANN: a library for approximate nearest neighbor searching (2010), http://www.cs.umd.edu/~mount/ANN/
13. C. Pacchierotti, D. Prattichizzo, K.J. Kuchenbecker, A data-driven approach to remote tactile interaction: from a biotac sensor to any fingetip cutaneous device, in *Haptics: Neuroscience, Devices, Modeling, and Applications. Eurohaptics 2014*. Lecture Notes in Computer Science, Versailles, France (2014), pp. 418–424
14. G.B. Limentani, M.C. Ringo, F. Ye, M.L. Bergquist, E.O. McSorley, Beyond the t-test: statistical equivalence testing. Anal. Chem. **77**(11), 221–226 (2005)
15. L. Meli, C. Pacchierotti, D. Prattichizzo, Sensory subtraction in robot-assisted surgery: fingertip skin deformation feedback to ensure safety and improve transparency in bimanual haptic interaction. IEEE Trans. Biomed. Eng. **61**(4), 1318–1327 (2014)
16. C. Chen, N. Rathore, W. Ji, A. Germansderfer, Statistical equivalence testing for assessing bench-scale cleanability. BioPharm Int. **23**, 2 (2010)
17. C. Lauzon, B. Caffo, Easy multiplicity control in equivalence testing using two one-sided tests. Am. Stat. **63**(2), 147–154 (2009)
18. C.L. Olson, On choosing a test statistic in multivariate analysis of variance. Psychol. Bull. **83**(4), 579 (1976)
19. A.V. Oppenheim, R.W. Schafer, J.R. Buck, et al., *Discrete-Time Signal Processing*, vol. 2 (Prentice-Hall Englewood Cliffs, Upper Saddle River, 1989)
20. T. Yamamoto, B. Vagvolgyi, K. Balaji, L.L. Whitcomb, A.M. Okamura, Tissue property stimation and graphical display for teleoperated robot-assisted surgery, in *Proceedings of the IEEE International Conference on Robotics and Automation* (2009), pp. 4239–4245
21. S.J. Lederman, R.L. Klatzky, Hand movements: a window into haptic object recognition. Cognitive Psychol. **19**(3), 342–368 (1987)

Part II
Force Feedback via Mixed Cutaneous and Kinesthetic Cues

Chapter 5
Cutaneous and Kinesthetic Cues to Improve Transparency in Teleoperation

Abstract The first part of the book showed promising results and proved the effectiveness of employing cutaneous-only feedback in various teleoperation scenarios. However, providing solely cutaneous stimuli still showed significantly worse performance than providing the user with full haptic feedback. This chapter takes up this challenge and proposes a new system able to combine cutaneous and kinesthetic feedbacks, with the objective of improving the transparency of teleoperation systems with force reflection while guaranteeing their safety. The proposed control algorithm integrates kinesthetic haptic feedback, provided by common grounded haptic interfaces, with cutaneous haptic feedback, provided by ungrounded cutaneous devices. The proposed approach can be used on top of any time-domain control technique that ensures a stable interaction by scaling down kinesthetic feedback when this is required to satisfy stability conditions (e.g., passivity) at the expense of transparency. Performance is then recovered by providing a suitable amount of cutaneous force through custom ungrounded cutaneous devices. The viability of the proposed approach is demonstrated through an experiment of perceived stiffness and an experiment of teleoperated needle insertion in soft tissue.

5.1 Introduction

As extensively discussed in Part I, haptic stimuli play a fundamental role in enhancing the performance of teleoperation systems. However, despite its expected benefit, teleoperation systems with haptic feedback are still missing from the market. One of the reasons for this omission is the negative effect that haptic feedback may have on the stability of these systems, as shown in the experiments of Chaps. 2 and 3.

This chapter is reprinted with kind permission from ACM, originally published in [1], and IEEE, originally published in [2].

C. Pacchierotti, *Cutaneous Haptic Feedback in Robotic Teleoperation*,
Springer Series on Touch and Haptic Systems, DOI 10.1007/978-3-319-25457-9_5

Guaranteeing the stability of this kind of teleoperation systems while preserving their transparency has thus always been a great challenge.

To this aim, researchers have proposed a great variety of transparency- and stability-optimized bilateral controllers [3, 4] and it has always been difficult to find a good trade-off between these two objectives. In this respect, passivity [5] has been exploited as the main tool for providing a sufficient condition for stable teleoperation in several controller design approaches such as the Scattering Algorithm [6], Time Domain Passivity Control [7], Energy Bounding Algorithm [8] and Passive Set Position Modulation [9]. In [6] a coding scheme is applied to the power variables (velocities and forces) to turn the time-delayed communication channel into a passive element. When the controllers at both the master and slave sides are, furthermore, passive, the complete system can be considered passive.

In [9], the authors propose an approach built around a spring-damper controller, where the energy dissipated by the *virtual* damper is stored in an energy tank and jumps in spring potential are limited to the available energy in the tank. More recently, a dual-layer controller structure has been presented by Franken et al. [10]. A transparency layer is in charge of computing the ideal forces to be actuated at both the master and the slave, regardless of passivity constraints. On the other hand, a passivity layer modulates such forces when this is necessary to avoid violations of the passivity condition, thus guaranteeing stability at the price of a temporary loss of transparency.

As discussed in Part I, a further approach to stability in teleoperation is *sensory substitution*. It consists of substituting haptic force with alternative forms of feedback, such as vibrotactile [11], auditory, and/or visual feedback [12]. In this case, since no kinesthetic force is fed back to the operator, the haptic loop becomes intrinsically stable and no bilateral controller is thus needed (see Chaps. 2 and 3). Our sensory subtraction approach described in Part I is similar to sensory substitution. It consists of substituting haptic feedback with cutaneous feedback only (see Fig. 1.1b). As discussed in Sect. 3.2.2, if the cutaneous devices are suitably designed, the force feedback minimally influences the position of the end-effector of the master device, thus opening the haptic loop and making the system intrinsically stable. Sensory subtraction shows significantly higher transparency levels compared to other conventional sensory substitution techniques, as shown in Chaps. 2 and 3. However, sensory subtraction usually provides the user with less transparency than that achieved using full haptic feedback [13–17].

In this chapter we present a novel technique based on the *combination* of kinesthetic and cutaneous feedback. It combines the promising cutaneous-only approach of sensory subtraction with the time-domain passivity control algorithm of [10], with the objective of obtaining better performance while guaranteeing the same degree of safety. The proposed approach provides force feedback through a grounded haptic device as long as the passivity condition is not violated. As the passivity controller detects a violation—and the haptic device is thus not guaranteed to provide the required feedback in a stable manner—kinesthetic feedback is scaled down and a cutaneous interface conveys a *suitable* amount of additional cutaneous force to recover transparency (see Fig. 5.1). The proposed strategy yields a teleoperation

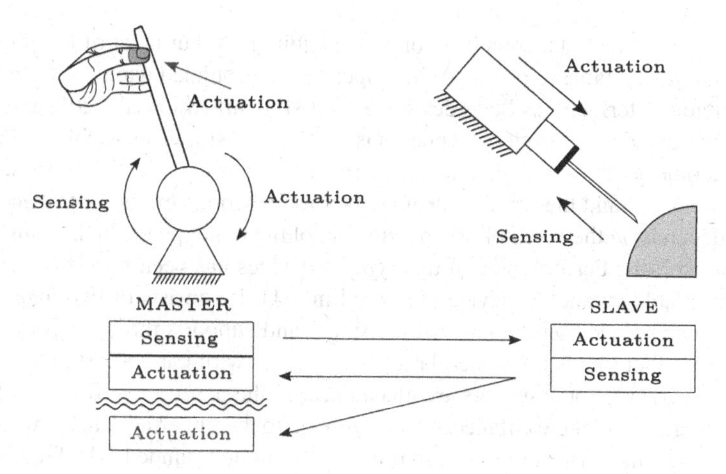

Fig. 5.1 Enhanced cutaneous-kineshetic approach proposed in this chapter. Force feedback on the master side is computed according to [10] and actuated via the grounded haptic device, as long as the passivity condition is not violated. As the passivity layer detects a violation, a cutaneous interface conveys additional cutaneous force to recover transparency. Actuation and sensing schemes for classic haptic teleoperation and sensory subtraction are shown in Fig. 1.1a and b, respectively

system which is stable due to passivity control, but with improved realism, since cutaneous feedback conveys force information that cannot be provided through the haptic interface.

The chapter is divided in two main parts. Section 5.3 discusses how to compute the *suitable* amount of cutaneous force necessary to recover from a loss of transparency due to a reduction of kinesthetic feedback. Section 5.4 presents the control algorithm, which uses the results of Sect. 5.3 to improve the transparency of passive teleoperation systems with force reflection.

The proposed approach is evaluated in two benchmark scenarios. In the first experiment we test the performance in terms of perceived stiffness of a virtual hard constraint using classic haptic feedback provided by a grounded haptic interface and the proposed cutaneous-kinesthetic approach. The second scenario involves a teleoperated needle insertion in soft tissue. Task performance is compared for the following cases: classic haptic feedback computed according to [10], cutaneous feedback only (sensory subtraction approach), and the proposed mixed method. Some preliminary results on this approach were presented in [18, 19].

5.2 An Ungrounded Fingertip Cutaneous Device

In order to improve the transparency of teleperation systems with force reflection, in this chapter we propose to provide cutaneous force *in addition* to the haptic feedback provided by grounded haptic interfaces. This requires the operator to use the end-effector of the grounded haptic device *in combination* with an interface able to provide the additional cutaneous force.

The literature on cutaneous technologies is quite rich, but most of the proposed devices are not suitable to be used while operating a grounded haptic device.

A suitable interface has been developed by Minamizawa et al. [20], where the authors presented an ungrounded cutaneous display that simulates weight sensations of virtual objects. However, this device cannot render forces in all directions, it has only two motors, and the force control is open-loop. Moreover, its control accuracy largely depends on the visco-elastic parameters of the finger pad, which change with different subjects. Performance of this type of devices has been improved with the 3-DoF wearable cutaneous device presented in [21]. It consists of two main parts: the first one is placed on the back of the finger and supports three small electrical motors; the other one is a mobile platform in contact with the volar skin surface of the fingertip. Three force sensors are placed near to the platform vertices, in contact with the finger, so that the cutaneous force applied to the fingertips can be measured.

The cutaneous device employed in this work is an ungrounded 3-DoF cutaneous interface similar to the one presented in [21] but showing higher accuracy and wearability. The device is shown in Fig. 5.2. It is composed of a static platform that houses three servo motors and a mobile platform that applies the requested stimuli to the fingertip. The two platforms are connected by three cables made of ultra-high-molecular-weight polyethylene [22]. Three small electrical motors, equipped with position encoders, control the length of the cables, to be able to move the platform toward the fingertip. One force sensor is placed at the platform's center, in contact with the finger. It has a diameter of 5 mm and a thickness of 0.3 mm, making it very transparent to the user and easy to integrate with the mobile platform. Its overall light weight, 35 g, and the small dimension of the mobile platform, make this cutaneous device suitable to be used together with common grounded haptic interfaces (see Fig. 5.3a and Chap. 2). The actuators used in our prototype are three 0615S DC micromotors (Dr. Fritz Faulhaber GmbH & Co., Germany), with planetary gear-heads having 16:1 reduction ratio. The maximum stall torque of the motor is 3.52 mNm. The encoders are 10-bit rotary position sensors named AS5040 (AMS AG, Austria) and the force sensors is a piezoresistive 400 FSR (Interlink Electronics, USA). With respect to the interface presented in [21], this device can be controlled both in force and position. Moreover, the smaller size improves its wearability.

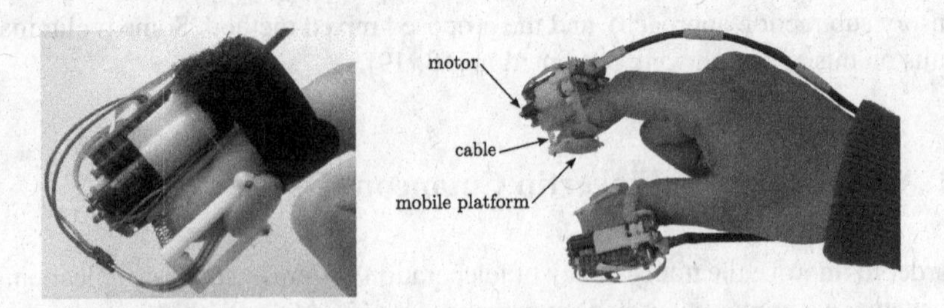

Fig. 5.2 The cutaneous device used in this work. It is similar to the one employed in Chap. 2, but it has improved performance, wearability and closed-loop force and position control

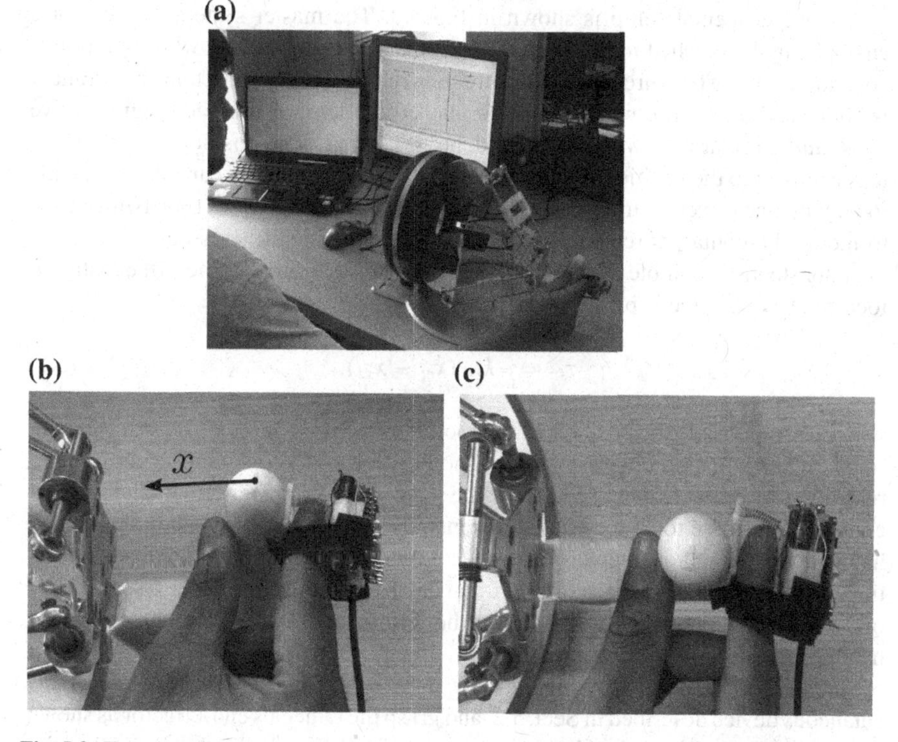

Fig. 5.3 Experimental setup. Users were asked to wear one cutaneous device on the index finger and grasp the handle, positioned with its longitudinal axis at 90° from the Omega x-axis. This setup is similar to the one described in Chap. 2. **a** User performing the 1-DoF teleoperation task. **b** *Front view* of the handle. **c** *Top view* of the handle

Note that, since the experiments described in this chapter consist of 1-DoF telemanipulation tasks, this device is used as a 1-DoF cutaneous device (all motors pulled the cables together), so that only the forces in the sagittal plane of the finger are actuated, roughly normal to the longitudinal axis of the distal phalanx (as in Chap. 2).

5.3 Compensating for a Reduction of Kinesthetic Feedback with Cutaneous Stimuli

In order to evaluate as to what extent acting on the stimuli provided by a cutaneous device can improve the transparency of teleoperation systems, we carried out two experiments. Their results will be used in Sect. 5.4 to design a controller able to improve the transparency of passive teleoperation systems by compensating any reduction of kinesthetic feedback with cutaneous force.

The experimental setup is shown in Fig. 5.3. The master system consists of a custom handle attached to the end-effector of an Omega 3 device, whose motion is constrained along one direction (the x direction in Fig. 5.3b). The virtual environment is composed of a stiff constraint, which plays the role of a forbidden-region active constraint, similarly to what we have done in Chap. 2. The position of the virtual tool is linked to the position of the haptic handle, so the operator can teleoperate the tool along one direction toward the stiff constraint. A spring $k_{sc} = 1800\,\text{N/m}$ is used to model the contact force τ_{sc} between the tool and the stiff constraint. When the operator steers the remote tool toward the unsafe area delimited by the stiff constraint, located at x_{sc}, a force is provided to the operator

$$\tau_{sc} = -k_{sc}\,(x_{rt} - x_{sc}), \tag{5.1}$$

where x_{rt} represents the position of the tool.

The Omega haptic device measures the position of the operator's hand (with a resolution of 0.01 mm) and sends it to the controller. Then the virtual environment computes the force to be fed back and transmits it to the user through the grounded haptic interface and cutaneous device, according to the considered feedback condition. To extend the workspace in the virtual environment, we introduce a scale factor of 3 between the position of the tool in the virtual environment and the operator's hand.

Subjects were asked to wear on their right index finger one prototype of the cutaneous device described in Sect. 5.2, and grasp the Omega's end-effector as shown in Fig. 5.3. The subject's hand was positioned with its longitudinal axis at 90° from the Omega x-axis. Before the beginning of each experiment we carefully checked the position of the hand with respect to the handle, the correct functioning of the device and of the virtual environment. To prevent changes in the perceived direction of the force feedback generated by the Omega, subjects were instructed to move the forearm, rather than the wrist, while moving the device. During the experiments, subjects maintained the initial orientation of the fingers with respect to the end-effector, which was the natural way of grasping the handle for the given 1-DoF task. As for the haptic rendering, the interaction was designed according to the god-object model [23] and the position of the Omega handle was linked to the tool position x_{rt} moving in the virtual environment. Subjects could see the moving tool on the screen, but they were not aware of the position of the stiff constraint.

Subjects were asked to move the remote tool across the virtual environment and stop the motion as soon as the stiff constraint was perceived (i.e., the users had to remain in contact with the stiff constraint). After 3 s of continuous contact with the constraint, the system played a beep sound. Subjects were instructed to move the tool back as soon as the auditory signal was heard. The position of the stiff constraint x_{sc} was chosen randomly at the beginning of each trial.

Sixteen participants (12 males, 4 females, age range 21–33) took part in the experiment, all of whom were right-handed. Seven of them had previous experience with haptic interfaces. None of the participants reported any deficiencies in their perception abilities.

5.3.1 Simple Compensation with Cutaneous Stimuli

The first experiment aims at evaluating the performance degradation rate when reducing the force feedback τ_h provided by the grounded haptic interface, and discusses to what extent providing additional cutaneous force τ_c through the cutaneous device can mitigate the consequent performance degradation.

Each participant made sixty trials of the 1-DoF teleoperation task described above:

- thirty trials with force feedback provided by *both* the cutaneous device and the grounded haptic interface, with three randomized repetitions for each feedback condition HC_α indicated in Table 5.1a, and
- thirty trials with force feedback provided by the grounded haptic interface *only*, with three randomized repetitions for each feedback condition H_α indicated in Table 5.1b.

Table 5.1 Simple compensation

Condition	α	β
(a) Compensation via cutaneous feedback, $\alpha + \beta = 1$		
HC_0	0	1
$HC_{0.1}$	0.1	0.9
$HC_{0.2}$	0.2	0.8
$HC_{0.3}$	0.3	0.7
$HC_{0.4}$	0.4	0.6
$HC_{0.5}$	0.5	0.5
$HC_{0.6}$	0.6	0.4
$HC_{0.7}$	0.7	0.3
$HC_{0.8}$	0.8	0.2
$HC_{0.9}$	0.9	0.1
(b) Force provided by the Omega 3 only, $\beta = 0$		
$H_{0.1}$	0.1	0
$H_{0.2}$	0.2	0
$H_{0.3}$	0.3	0
$H_{0.4}$	0.4	0
$H_{0.5}$	0.5	0
$H_{0.6}$	0.6	0
$H_{0.7}$	0.7	0
$H_{0.8}$	0.8	0
$H_{0.9}$	0.9	0
H_1	1	0

Variables α and β represent the portion of force provided, respectively, by the grounded haptic interface and the cutaneous device. In condition $HC_{0.2}$, for example, the force to be applied at the master side, τ_{sc}, is provided by the Omega 3 and the cutaneous device scaled to the 20 % and 80 %, respectively. On the other hand, in condition $H_{0.2}$, the force τ_{sc} is provided through the Omega only, scaled to the 20 %

Subscript α refers to the portion of the total force feedback τ_{sc} provided through the grounded haptic interface, $\alpha = \frac{\tau_h}{\tau_{sc}}$. Similarly, we can define the portion of the total force feedback provided through the cutaneous device as $\beta = \frac{\tau_c}{\tau_{sc}}$ (see Table 5.1). Force τ_{sc} is calculated according to Eq. (5.1). In conditions HC_α the force is fed back by both the cutaneous device and the Omega, and the cutaneous device compensates for any reduction of the force provided by the grounded interface, i.e., $\alpha + \beta = 1$. In conditions H_α the force is fed back through the Omega device only, i.e., $\beta = 0$. Condition H_1, when all force is fed back through the Omega interface and the cutaneous device is not active, is considered our control condition, i.e., the one from which we expect the best performance.

Subjects were asked to wear the cutaneous device on their right index finger, grasp the end-effector of the Omega as shown in Fig. 5.3, and complete the sixty aforementioned trials (conditions HC_α and H_α were mixed together). Each subject was informed about the procedure before the beginning of the experiment, and a 10-min familiarization period was provided to acquaint the participants with the experimental setup. Since subjects were asked to stop as soon as the virtual stiff constraint was perceived, the average penetration inside the stiff constraint provided a measure of accuracy in reaching the target depth. A null value in the metrics denoted the best performance, while a positive value indicated that the subject overran the target.

Figure 5.4a shows the average penetrations, beyond the stiff constraint, for each feedback condition HC_α (circles) and H_α (squares). Figure 5.4b shows the improvement, in terms of penetration beyond the stiff constraint, from H_α to HC_α, with respect to the portion of force provided by the grounded interface. The collected data of each condition passed the D'Agostino-Pearson omnibus K2 normality test. We ran nine paired t-tests to evaluate the statistical significance of the difference between conditions H_α and HC_α that provide the same portion of force through the grounded interface, e.g., $H_{0.1}$ versus $HC_{0.1}$, $H_{0.2}$ versus $HC_{0.2}$, etc. Results revealed statistically significant difference for conditions with $\alpha \leq 0.6$ (depicted as filled markers

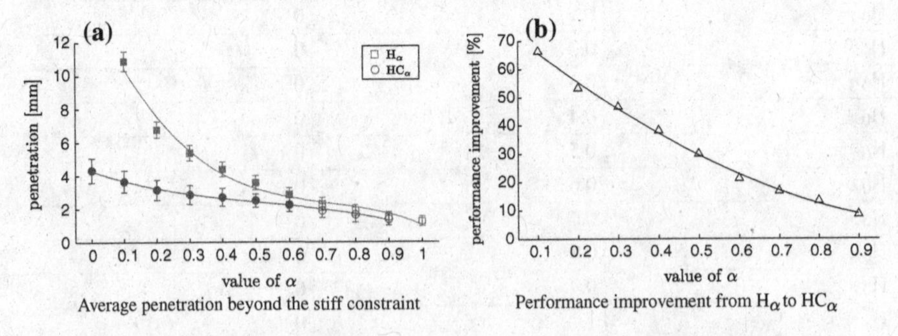

Fig. 5.4 Simple compensation. Average penetration beyond the stiff constraint (mean and standard deviation) and performance improvement from HC_α to H_α, in terms of penetration beyond the stiff constraint. *Filled markers* in (**a**) represent the conditions found statistically different. *Solid lines* represent the cubic approximation to the data sets

in Fig. 5.4a). However, also in conditions whose results were not found significantly different ($\alpha > 0.6$), subjects still showed better performances when receiving additional cutaneous feedback from the cutaneous device.

Results indicate that the subjects, while receiving force feedback through the Omega only (conditions H_α), reached a greater average penetration in the stiff constraint (worse performance) then that obtained while receiving feedback *also* from the cutaneous device (conditions HC_α). This is, of course, more evident as the value of α decreases.

No difference between conditions was observed in terms of task completion time.

The observed results are in agreement with previous findings. In Chap. 2 we found an average penetration inside the constraint of ~1.7 mm for the feedback condition called here H_1, and of ~3.9 mm for HC_0. Moreover, it is also interesting to notice how users, during conditions H_α, tended to stop when the force exerted by the Omega interface reached a certain reference value (~2.5 N), regardless of the penetration inside the stiff constraint. During conditions HC_α, as expected, this reference force decreased, thanks to the additional cutaneous stimuli provided by the cutaneous device.

5.3.2 Over-Compensation with Cutaneous Stimuli

The second experiment aims at evaluating the extent to which an *over*-compensation of a reduction of kinesthetic feedback through cutaneous stimuli can improve the results registered in the previous section. The experimental protocol is similar to the one described before, but this time we consider conditions with $\alpha + \beta > 1$.

Each participant made thirty-six trials of the same 1-DoF teleoperation task presented in Sect. 5.3.1. Force feedback was provided by *both* the cutaneous device and the Omega, with two randomized repetitions for each feedback condition $HCO_{\alpha,\beta}$ shown in Table 5.2. These trials were performed right after the ones described in Sect. 5.3.1, and the participants were not aware that they were performing a different experiment. The average penetration inside the stiff constraint provided again a measure of accuracy in reaching the target depth.

Figure 5.5a shows the average penetration beyond the stiff constraint for feedback conditions $HCO_{\alpha,\beta}$, with respect to $\alpha + \beta$. Different lines indicate different levels of α. Figure 5.5b shows the average penetration beyond the stiff constraint for the same feedback conditions, but with respect to α. Different lines indicate different levels of $\alpha + \beta$. The dashed black line represents the performance of our control condition H_1, i.e., when all the force was fed back through the grounded interface. The collected data of each condition passed the D'Agostino-Pearson omnibus K2 normality test. In order to determine whether the results observed can be considered as statistically equivalent to our control condition H_1, we performed a two one-sided t-test (TOST). The null hypothesis of the TOST states that the mean values of two groups are different by a certain amount ε (or larger). The design of equivalence tests

Table 5.2 Over-compensation

Condition	α	β
$HCO_{0.4,0.8}$	0.4	0.8
$HCO_{0.4,1.0}$	0.4	1.0
$HCO_{0.4,1.2}$	0.4	1.2
$HCO_{0.5,0.7}$	0.5	0.7
$HCO_{0.5,0.9}$	0.5	0.9
$HCO_{0.5,1.1}$	0.5	1.1
$HCO_{0.6,0.6}$	0.6	0.6
$HCO_{0.6,0.8}$	0.6	0.8
$HCO_{0.6,1.0}$	0.6	1.0
$HCO_{0.7,0.5}$	0.7	0.5
$HCO_{0.7,0.7}$	0.7	0.7
$HCO_{0.7,0.9}$	0.7	0.9
$HCO_{0.8,0.5}$	0.8	0.4
$HCO_{0.8,0.7}$	0.8	0.6
$HCO_{0.8,0.9}$	0.8	0.8
$HCO_{0.9,0.5}$	0.9	0.3
$HCO_{0.9,0.7}$	0.9	0.5
$HCO_{0.9,0.9}$	0.9	0.7

Each participant made thirty-six randomized trials of the teleoperation task, in addition to the one already performed in Sect. 5.3.1, with two randomized repetitions for each feedback condition shown in this table. Again, variables α and β represent the portion of force provided, respectively, by the grounded haptic interface and the cutaneous device

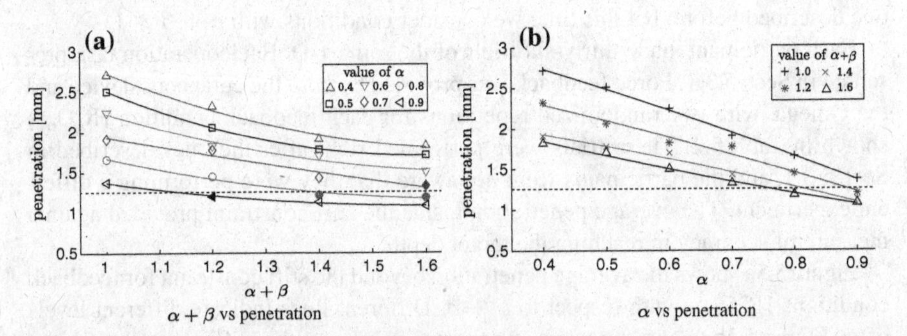

Fig. 5.5 Over-compensation. Average penetration beyond the stiff constraint (mean and standard deviation) with respect to $\alpha + \beta$ and α. A null value of this metric indicates high accuracy in reaching the target depth. *Filled markers* in (**a**) represent the conditions found statistically equivalent to the control condition H_1 (*dashed black line*). *Solid lines* represent the cubic approximation to the data sets

can be quite tricky due to the fact that the acceptance criterion ε has to be defined on the basis of prior knowledge of the measurement. Similar to Sect. 4.2.3.1, we again evaluated the acceptance criterion ε as suggested in [24].

A two one-sided t-test was performed between H_1 and *each* $HCO_{\alpha,\beta}$ condition shown in Table 5.2. In order to avoid raising the family-wise error rate, i.e., the probability of at least one incorrectly rejected null in a family of tests, we took into account the simple correction presented by Lauzon and Caffo [25]. The tests revealed statistical equivalence between H_1 and $H_{0.7,0.9}$, $H_{0.8,0.8}$, $H_{0.8,0.6}$, and $H_{0.9,0.3}$ (depicted as filled markers in Fig. 5.5a). Moreover, we tested for difference the means along the α and $\alpha + \beta$ dimensions using a two-way ANOVA. The means differed significantly for the portion of force provided by the grounded interface ($F_{5,360} = 104.25$, $p < 0.001$) and the total force ($F_{3,360} = 85.370$, $p < 0.001$). Also the interaction between the factors was found significant ($F_{15,360} = 2.210$, $p = 0.006$). Regarding the portion of force provided by the grounded interface, α, post-hoc analysis (pair-wise post-hoc t-tests corrected according to Bonferroni) revealed statistically significant difference for all the conditions. Regarding the total force, $\alpha + \beta$, post-hoc analysis (pair-wise post-hoc t-tests corrected according to Bonferroni) revealed statistically significant difference for all the conditions but for $\alpha + \beta = 1.4$ versus $\alpha + \beta = 1.6$. This may be due to the limited actuation capabilities of the cutaneous device, which is not able to exert more than 2.5 N at the fingertip.

The results of this experiment indicate that subjects reached smaller average penetration in the stiff constraint (better performance) when over-compensating ($\alpha + \beta > 1$) with respect to the simple compensation presented in Sect. 5.3.1 ($\alpha + \beta = 1$). Moreover, in some of the considered conditions, the over-compensation via cutaneous stimuli lead to performance comparable to that obtained while employing solely the grounded haptic interface ($\alpha = 1$ and $\beta = 0$, our best condition H_1). Results therefore indicate that, to a certain extent, it is possible to compensate for a reduction of force feedback with cutaneous stimuli, expecting no significant performance degradation. However, it is worth noting that the effectiveness of the proposed approach is directly related to the effectiveness of the cutaneous device and to the nature of the task.

5.3.2.1 A More General Approach

We have analyzed the performance of the system while compensating via cutaneous stimuli up to certain levels, i.e., $\alpha + \beta = 1.2, 1.4, 1.6$. However, it may be useful to find a more general law that indicate the level of cutaneous stimuli needed to compensate for any given reduction of the force feedback provided by the grounded interface. From data shown in Fig. 5.5 it is indeed possible to evaluate a polynomial fitting of performance versus actuation of the grounded interface, and then find the intersection between this curve and the reference value H_1. From that, we can evaluate the function

$$\beta = g(\alpha), \tag{5.2}$$

which can be expanded as

$$\tau_c = g\left(\frac{\tau_h}{\tau_{sc}}\right)\tau_{sc}, \tag{5.3}$$

which gives us the amount of cutaneous force to be provided through the cutaneous interface, τ_c, to fully compensate for a given reduction of force feedback $\frac{\tau_h}{\tau_{sc}}$.

However, it is important to notice that it is not possible to compensate for *any* reduction of force feedback through this technique. This is mainly due to the limited capability of the cutaneous device in conveying sufficient information to the human user. In fact, our $g(\alpha)$ is found here to be defined only for $1 \geq \alpha \geq \alpha_{min} \approx 0.75$. Below α_{min} it is not possible to *fully* recover transparency, but only to mitigate the degradation of performance by conveying as much force as possible through the cutaneous device.

This approach will be used in the next section to design a passivity algorithm that guarantees the same stability properties of common passivity approaches but improves their transparency. It aims at compensating any reduction of force feedback due to stability constraints with additional cutaneous stimuli, since cutaneous force does not affect the stability of teleoperation systems. The cutaneous stimuli to be provided will be computed as in Eq. (5.3).

5.4 Enhancing the Performance of Passive Teleoperation Systems with Cutaneous Stimuli

In this section we discuss how to integrate the sensory subtraction technique of Part I, the results presented in Sect. 5.3, and the passivity-based controller of [10]. The objective is to design a novel passivity algorithm that guarantees the same stability properties of [10] but significantly improves its transparency.

5.4.1 *Time-Domain Passivity Control for Haptic Force Feedback*

The passivity-based time-domain control scheme by Franken et al. [10] guarantees a stable behavior of bilateral telemanipulation systems in the presence of time-varying destabilizing factors, such as hard contacts, relaxed user grasps, stiff control settings, and/or communication delays. The architecture is split into two separate layers. The hierarchical top layer, named *Transparency Layer*, aims at achieving the desired transparency, while the lower layer, named *Passivity Layer*, ensures the passivity of the system (see Fig. 5.6a). The operator and the environment impress a movement q_m and q_s to the master and slave systems, respectively. The Transparency Layer displays the desired behavior to obtain transparency by computing the torques τ_{TLm} and τ_{TLs} to be applied to the operator and to the environment, respectively. The Passivity Layer checks how the action planned by the Transparency Layer influences the energy balance of the system. If the passivity condition is not violated, the planned action τ_{TL*} can be directly applied to both sides of the system. However, if loss of passivity

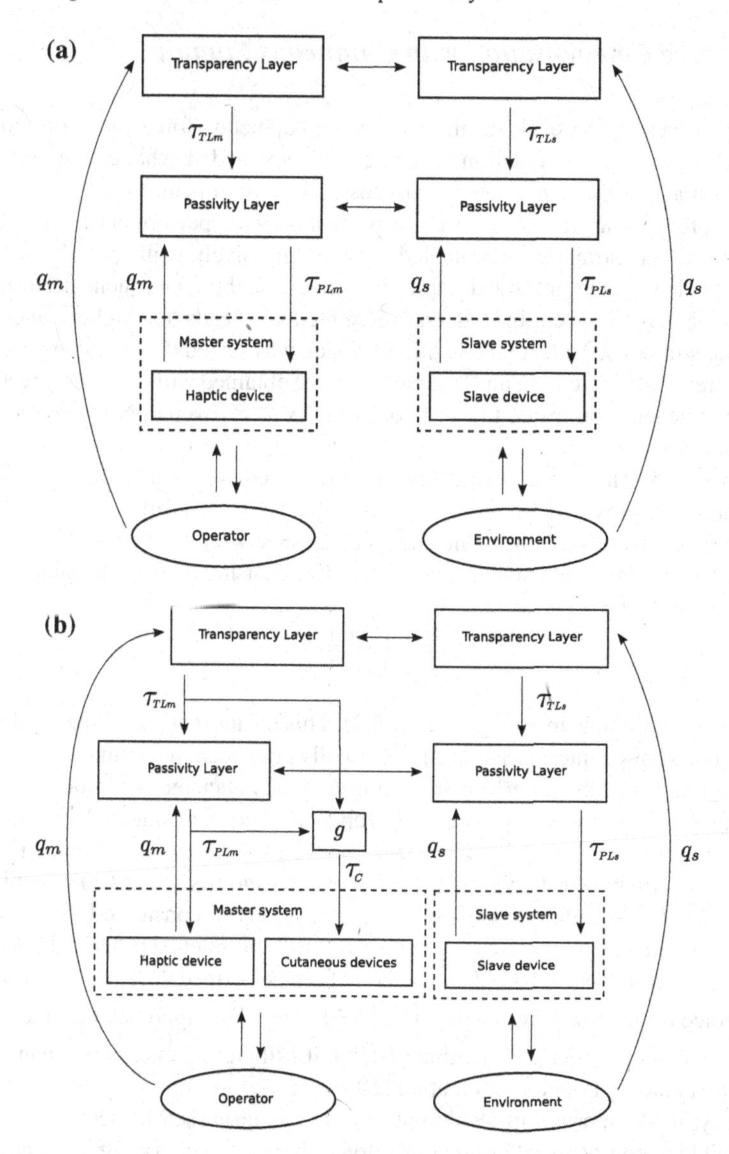

Fig. 5.6 Our approach modifies the control strategy in [10] by adding the opportunity of providing cutaneous feedback when the required force cannot be conveyed using kinesthesia. **a** Original approach presented by Franken et al. [10]. **b** The mixed kinesthetic-cutaneous feedback approach presented in this work

is detected, a scaled control action τ_{PL*} is applied to preserve stability, resulting in a loss of transparency. Separate communication channels connect the layers at the slave and master levels so that information related to exchanged energy is separated from information about the desired behavior.

5.4.2 Force Compensation with Cutaneous Stimuli

It is now necessary to evaluate the amount of cutaneous force to be provided to compensate for a given reduction of haptic feedback, and to what extent cutaneous stimuli can actually compensate for this loss. The experimental work discussed in Sect. 5.3 provides an insight into these problems. Task performance (penetration inside the stiff constraint) was evaluated while progressively scaling down the haptic force provided by the grounded haptic interface and the consequent performance degradation was analyzed. Indeed, less force feedback leads to a higher penetration inside the stiff constraint. As the haptic feedback was scaled down, cutaneous force was progressively increased, until the performance obtained with full haptic feedback (i.e., same penetration inside the stiff constraint) was recovered. No stability issues were considered.

As in Sect. 5.3, let us refer to the force to be rendered as τ_{sc} and to the scaled haptic force feedback provided by the grounded interface as τ_h (with $|\tau_h| \leq |\tau_{sc}|$). The additional cutaneous force for which the performance with cutaneous compensation was statistically equivalent to the one registered when using only the grounded device was estimated to be

$$\tau_c = g\left(\frac{\tau_h}{\tau_{sc}}\right)\tau_{sc}, \tag{5.4}$$

where $g(\cdot)$ is a suitable mapping (see Eq. 5.3). This means that providing τ_{sc} through the grounded haptic interface showed statistically equivalent performance as providing τ_h through the same interface and τ_c through the cutaneous actuator.

Using this experimental protocol, a proper $g(\cdot)$ can be evaluated for any teleoperation scenario. Note that $g(\cdot)$ is task- as well as device-dependent. In all the experiments conducted, however, it turned out that $g(\cdot)$ is strictly monotonic and greater than $1 - \frac{\tau_h}{\tau_{sc}}$, regardless of the particular scenario considered. Evaluating a proper $g(\cdot)$ for a given scenario may require a long experimental process. In Sect. 5.3 data was gathered from 16 subjects, each of whom performed 96 trials. A quick-and-dirty choice for $g(\cdot)$ may simply be $g\left(\frac{\tau_h}{\tau_{sc}}\right) = 1 - \frac{\tau_h}{\tau_{sc}}$. This approach provides worse performance than properly evaluating $g(\cdot)$, but it still yields better performance than using no cutaneous compensation at all [19].

Finally, it is important to also point out that, as discussed in Sect. 5.3.2.1, it is not possible to compensate for *any* reduction of haptic force through this technique. This is mainly due to the limited capability of cutaneous stimulation and to the technological limitations of the cutaneous actuator employed. Over a certain value of $|\tau_{sc}(k) - \tau_h(k)|$ it is not possible to compensate for the loss, but only to mitigate any degradation of performance by conveying as much force as possible through the cutaneous actuator.

In this work, proper mapping functions for the two experimental scenarios in Sects. 5.5.1 and 5.5.2 were evaluated following the protocol described in Sect. 5.3.

5.4.3 Enhanced Cutaneous-Kinesthetic Control Algorithm

With reference to Fig. 5.6b, the Transparency Layer evaluates the desired force feedback τ_{TLm} to be provided at the master side, while the Passivity Layer checks how the planned action influences the energy balance of the system. If the passivity condition is not violated, then τ_{TLm} can be fully applied to the operator through the grounded haptic interface. However, if loss of passivity is detected, only a scaled control action τ_{PLm}, such that $|\tau_{PLm}| < |\tau_{TLm}|$, can be applied through the grounded interface, in order to guarantee stability. In this case, we provide an additional amount of cutaneous force τ_c according to the method discussed in Sect. 5.3. We therefore set

$$\tau_c = g\left(\frac{\tau_{PLm}}{\tau_{TLm}}\right) \tau_{TLm},$$

where τ_{PLm} and τ_c are provided, respectively, through a grounded haptic device and the cutaneous actuator. The total force applied to the operator then turns out to be

$$\tau_t = \tau_{PLm} + \tau_c. \tag{5.5}$$

If no violation of the passivity conditions is detected, we have $\tau_c = 0$. Hence,

$$\tau_t = \tau_{PLm} + \tau_c = \tau_{TLm},$$

which means that all the force is provided through the grounded device (which is the ideal condition). We remark that $g(\cdot)$ is task- and device-dependent and should be newly evaluated for any different experimental setup according to the guidelines discussed in Sect. 5.3.

5.5 Experimental Evaluation

5.5.1 Experiment #1: Perceived Stiffness

In order to demonstrate the feasibility and effectiveness of our method, we carried out two experiments. The first one evaluates the perceived stiffness of a virtual environment, comparing the performance of the unaltered algorithm of Franken et al. [10] with our mixed approach.

5.5.1.1 Experimental Setup and Methods

The experimental setup is shown in Fig. 5.7. The master system is composed of two Omega 3 haptic interfaces and one prototype of the cutaneous device presented in

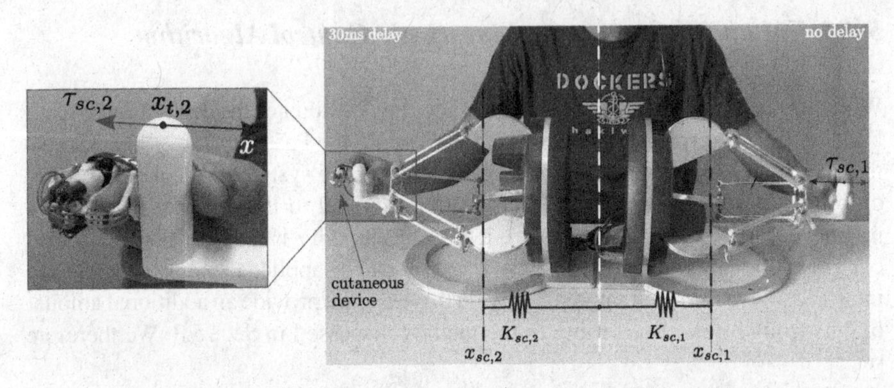

Fig. 5.7 Experiment #1. The master system is composed of two Omega haptic interfaces $n = 1, 2$ and one cutaneous device. A simulated master-slave communication delay of 30 ms was introduced between the Omega controlled by the subject's right hand and its stiff constraint

Sect. 5.2. Subjects are asked to wear one cutaneous device on the right index finger, and grasp the Omega's end-effectors as shown in Fig. 5.7. The motion of the Omega interfaces is constrained along the x-axis. Each interface interacts with a virtual stiff constraint. When subjects steer a haptic interface toward the workspace area delimited by its stiff constraint, the system computes the ideal force to be fed back as

$$\tau_{sc,n} = k_{sc,n}(x_{t,n} - x_{sc,n}), \quad n = 1, 2 \tag{5.6}$$

where $x_{t,n}$ indicates the position of the nth interface, while $x_{sc,n}$ and $k_{sc,n}$ indicate the position and stiffness of the nth constraint, respectively.

In order to highlight the role of our cutaneous compensation technique, a simulated master-slave communication delay of 30 ms is introduced between the 2nd Omega and its virtual environment (left in Fig. 5.7). This delay brings the system close to instability as stiffness increases. On the contrary, no delay is introduced between the 1st Omega and its virtual environment. This fact, combined with a high sampling rate (∼7 kHz), prevents the 1st Omega from showing any unstable behavior for the employed stiffness values.

The 1st Omega (on the right in Fig. 5.7), when in contact with the stiff constraint, always feeds back the ideal force $\tau_{sc,1}$. The 2nd Omega (on the left in Fig. 5.7) is equipped with a cutaneous device and can operate according to one of the two following feedback conditions:

- force feedback provided by the Omega only, as computed by the unaltered algorithm of [10] (condition F),
- force feedback provided by the Omega and the cutaneous device, as computed by the method presented in Sect. 5.4.3 (condition EF).

In condition F, if the passivity condition is not violated, the planned force $\tau_{PLm} = \tau_{TLm} = \tau_{sc,2}$ is directly fed back to the human subject via the Omega. Conversely, if loss of passivity is detected, a scaled action τ_{PLm} is applied. The cutaneous device is never active.

In condition EF, if the passivity condition is not violated, the planned force $\tau_{PLm} = \tau_{TLm} = \tau_{sc,2}$ is directly fed back to the human subject via the Omega, as in condition F. Conversely, if loss of passivity is detected, a scaled control action τ_{PLm} is applied through the Omega, and the cutaneous device provides the additional cutaneous force

$$\tau_c = g_1\left(\frac{\tau_{PLm}}{\tau_{TLm}}\right)\tau_{TLm}, \tag{5.7}$$

where $g_1(\cdot)$ is the mapping function that indicates the level of cutaneous stimuli needed to compensate for a reduction of haptic force during the considered task. Function $g_1(\cdot)$, shown in Fig. 5.8, has been evaluated for this task according to the guidelines described in Sect. 5.3.

We tested the perceived stiffness of the virtual environment for reference values of stiffness $k_{sc,ref}$ between 250 N/m and 3000 N/m, with a step size of 250 N/m (12 values in total, see Fig. 5.9). For each reference value of stiffness $k_{sc,ref}$ and for each feedback condition, we evaluated the stiffness *perceived* by the subjects.

Each evaluation started by setting $k_{sc,1} \ll k_{sc,2} = k_{sc,ref}$. Subjects were asked to interact with the two stiff constraints and tell the experimenter which one *felt* stiffer. At first, as expected, all the subjects reported $k_{sc,2}$ to feel stiffer than $k_{sc,1}$. We then increased $k_{sc,1}$ by a fixed step size of 50 N/m and asked the subject again, and we kept increasing $k_{sc,1}$ until the subject reported $k_{sc,1}$ to feel stiffer than $k_{sc,2}$. At that point we took the average between the two last tested values of $k_{sc,1}$ as the perceived stiffness for the considered subject, reference stiffness, and feedback condition. In an ideal scenario (no stability issues), both Omega interfaces would accurately render

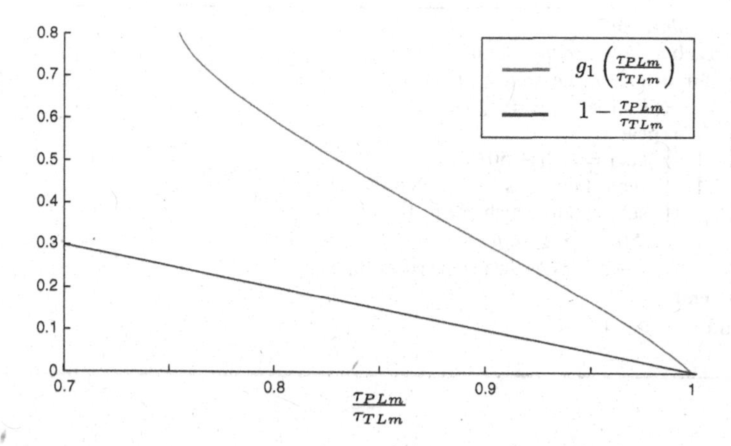

Fig. 5.8 Experiment #1. Function $g_1(\cdot)$ indicates the level of cutaneous stimuli needed to compensate for a certain reduction of haptic force

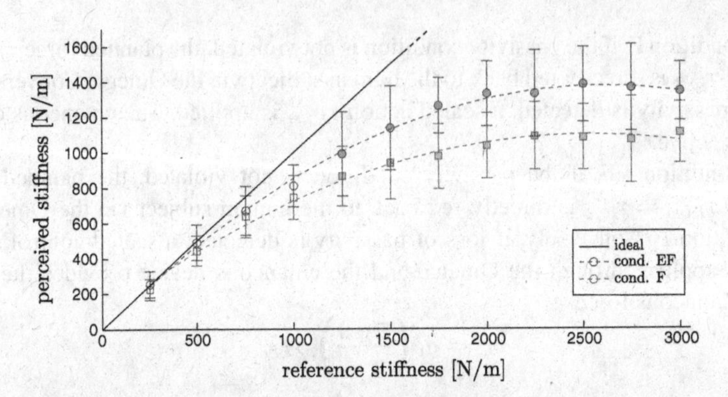

Fig. 5.9 Experiment #1. Average stiffness perceived by the subjects for the two feedback conditions and the twelve reference stiffness values. *Filled markers* show the stiffness values where the two feedback conditions were found statistically different from each other. *Dashed lines* represent the quadratic approximation to the data sets. The *black line* represents the ideal perceived stiffness

the stiffness of the respective constraints and, therefore, the perceived stiffness would always be very close to $k_{sc,ref}$. However, when the Passivity Layer needs to reduce the force feedback given by the 2nd Omega to guarantee the stability of the interaction, the constraint will feel less stiff than it should be. In this latter case, the perceived stiffness will be less stiff than $k_{sc,ref}$. The additional cutaneous force conveyed by the cutaneous device in condition EF aims at recovering this reduction of force. We expect subjects to perceive the stiff constraint *stiffer* when employing our enhanced mixed cutaneous-kinesthetic control approach with respect to the unaltered algorithm of [10]. For the sake of clarity, the experimental protocol has been summarized below.

Algorithm 1: Perceived stiffness experiment

foreach *subject* **do**
  foreach *feedback condition* **do**
    foreach *reference value of stiffness* $k_{sc,ref}$ **do**
      set $k_{sc,1} \ll k_{sc,2} = k_{sc,ref}$;
      repeat
        $k_{sc,1} = k_{sc,1} + 50\,\text{N/m}$;
        subject interacts with stiff constraints;
        subject tells which one feels stiffer;
      until ($k_{sc,1}$ *feels stiffer than* $k_{sc,2}$);
      $k_{sc,1} - 25\,\text{N/m}$ is the perceived stiffness;
    end
  end
end

Fifteen participants (13 males, 2 females, age range 20–29 years) took part in the experiment, all of whom were right-handed. Eight of them had previous experience with haptic interfaces. None reported any deficiencies in their perception abilities. Participants were informed about the procedure before the beginning of the experiment, and a 10-min familiarization period was provided to acquaint them with the experimental setup.

5.5.1.2 Results

In order to compare the performance of the two feedback conditions considered, we evaluated the perceived stiffness for the twelve reference values. A perceived stiffness lower than the ideal one indicated a loss of transparency in the system.

Figure 5.9 shows the average stiffness perceived by the subjects for the two feedback conditions and the twelve reference stiffness values. In order to determine whether the data registered differ between the two conditions, we ran twelve Wilcoxon signed-rank tests (significance level alpha = 0.05), one for each reference stiffness, i.e., F versus EF for $k_{sc,ref} = 250, 500, 750, \ldots, 3000$ N/m. The Wilcoxon signed-rank test is the non-parametric equivalent of the more popular paired t-test. The latter is not appropriate here since the dependent variable was measured at the ordinal level. The analysis revealed significant statistical difference between conditions F and EF for $k_{sc,ref} \geq 1250$ N/m (depicted as filled markers in Fig. 5.9). However, also when results were not found significantly different ($k_{sc,ref} < 1250$ N/m), subjects still reported better performances when receiving additional cutaneous force feedback by the cutaneous device. Details on the statistical analysis are reported in Table 5.3.

Table 5.3 Statistical analysis results for Experiment #1

Wilcoxon signed-rank test (EF – F, alpha = 0.05)		
$k_{sc,ref}$ (N/m)	Z statistic	p-values
250	−1.179	0.238
500	−1.941	0.052
750	−1.232	0.218
1000	−1.854	0.064
1250	−2.150	**0.032**
1500	−2.868	**0.004**
1750	−3.425	**0.001**
2000	−3.346	**0.001**
2250	−3.098	**0.002**
2500	−3.279	**0.001**
2750	−3.279	**0.001**
3000	−3.140	**0.002**

Z statistics are based on negative ranks. Bold p-values indicate significant difference

5.5.2 Experiment #2: Teleoperated Needle Insertion in Soft Tissue

The second experiment aims at evaluating the performance of our cutaneous-kinesthetic approach in a paradigmatic 1-DoF teleoperation experiment of needle insertion in soft tissue. In this experiment, we compared the performance while employing the unaltered algorithm of [10], the cutaneous-only sensory subtraction approach presented in Part I, and the proposed mixed cutaneous-kinesthetic method.

5.5.2.1 Experimental Setup and Methods

The experimental setup is shown in Fig. 5.10. The master system is composed of one Omega 3 haptic interface and two prototypes of the cutaneous device presented in Sect. 5.2. The motion of the Omega is constrained along its x-axis. Subjects wear one cutaneous device on the index finger, one cutaneous device on the thumb, and grasp the Omega's end-effectors as shown in Fig. 5.10a. The slave system is composed of a 6 DoF manipulator KUKA KR3 (KUKA Roboter GmbH, Germany), a 1-DoF force sensor, and a hypodermic needle, as shown in Fig. 5.10b. The needle is attached to the force sensor that, in turn, is attached to the end-effector of the KUKA manipulator. The needle, made of stainless-steel, has a diameter of 1 mm and a bevel angle at the tip of 30°. The environment is composed of a soft-tissue phantom made of gelatine mixture. A stiff object, made of polystyrene foam, is placed at 2 cm from the insertion point.

Subjects, through the haptic interface, control the motion of the slave robot. The force sensor registers the force τ_s exerted by the remote environment on the needle. According to the feedback condition being considered, the Omega 3 and the

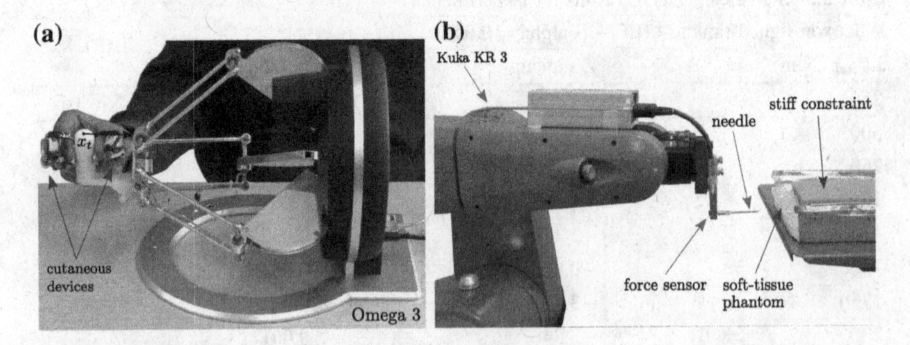

Fig. 5.10 Experiment #2. The master system is composed of one Omega haptic interface and two prototypes of the cutaneous device presented in Sect. 5.2. The motion of the Omega was constrained along is x-axis. The slave system is composed of a 6 DoF manipulator KUKA KR3, a 1-DoF force sensor, and a hypodermic needle. The needle is attached to a force sensor that, in turn, is fixed to the end-effector of the robotic manipulator. The environment is composed of a soft-tissue phantom made of gelatine mixture. A stiff object is placed 2 cm away from the insertion point. **a** Master system. **b** Slave system and environment

cutaneous devices feed back a suitable amount of force to the human subject. The task consisted in inserting the needle into the soft-tissue phantom and stopping the motion as soon as the stiff object was perceived. After 3 s of continuous contact with the object, the system played a beep sound. Subjects were instructed to pull the needle out of the soft-tissue phantom when the sound was heard. A video of the experiment is available as supplemental material at http://extras.springer.com/978-3-319-25455-5 and at http://goo.gl/cp0Nro.

Each participant made twelve randomized trials of the needle insertion task, with four repetitions for each feedback condition proposed:

- force feedback provided by the Omega only, as computed by the unaltered algorithm of [10] (condition F),
- force feedback provided by the cutaneous devices only, as in the sensory subtraction approach presented in Part I (condition C),
- force feedback provided by the Omega 3 and the cutaneous devices, as computed by the proposed mixed cutaneous-kinesthetic method (condition EF).

Condition F is the same as described in Sect. 5.5.1. The Transparency Layer is in charge of evaluating the ideal force to be provided, that in this case is the one registered by the force sensor at the slave side, hence $\tau_{TLm} = \tau_s$. If the passivity condition is not violated, then the planned force $\tau_{PLm} = \tau_{TLm} = \tau_s$ is applied to the human subject via the Omega device, otherwise a scaled force τ_{PLm} is provided. The cutaneous actuators are not active.

In condition C, the force τ_s registered by the force sensor is fed back through the cutaneous devices. The Omega interface only tracks the position of the fingers and does not provide any force. This is the sensory subtraction approach presented in Part I.

Condition EF is similar to the one described in Sect. 5.5.1. In case of violation of the passivity condition, a scaled force τ_{PLm} is provided through the Omega, while the cutaneous actuators provide

$$\tau_c = g_2\left(\frac{\tau_{PLm}}{\tau_{TLm}}\right)\tau_{TLm}, \tag{5.8}$$

where $g_2(\cdot)$ is the mapping function indicating the level of cutaneous stimuli needed to fully compensate for a reduction of haptic force during the considered task. Function $g_2(\cdot)$, shown in Fig. 5.11, has been evaluated for this task according to the guidelines described in Sect. 5.3.

In conditions C and EF, a positive cutaneous force directed toward the negative direction of the x-axis (see Fig. 5.10a) is provided by applying a normal stress to the index finger. Conversely, a negative cutaneous force, directed toward the positive direction of the x-axis is provided by applying a normal stress to the thumb. In all the considered conditions no visual feedback on the needle is provided.

Twenty participants (16 males, 4 females, age range 23–32 years) took part in the experiment, all of whom were right-handed. Four of them had previous experience with haptic interfaces. None reported any deficiencies in their perception abilities and

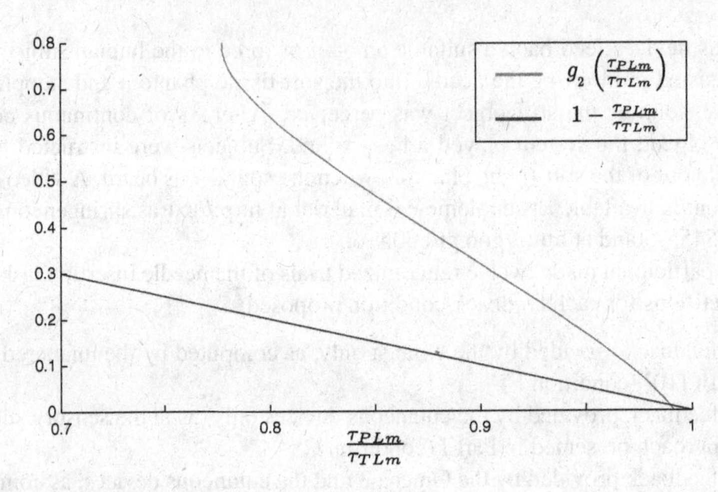

Fig. 5.11 Experiment #2. Function $g_2(\cdot)$ indicates the level of cutaneous stimuli needed to compensate for a certain reduction of haptic force

they were all naïve as to the purpose of the study. Participants were informed about the procedure before the beginning of the experiment, and a 10-min familiarization period was provided to acquaint them with the experimental setup.

5.5.2.2 Results

With the aim of comparing the performance of the three different feedback conditions, we evaluated the average needle penetration inside the stiff constraint, the maximum needle penetration inside the stiff constraint, and the average force reduction due to passivity constraints, computed as the mean over time of $\tau_{TLm} - \tau_{PLm}$. Data resulting from different trials of the same condition, performed by the same subject, were averaged before comparison with other conditions. Such metrics provide a measure of accuracy (average penetration), overshoot (maximum penetration), and transparency (force reduction) for the given task. Penetration measures can be considered particularly relevant to the medical scenario, as an excessive penetration of the needle can result in permanent damage of tissues. Moreover, a high force reduction severely compromises the realism of the haptic interaction.

Figure 5.12 shows the trajectory of the needle (solid red line) versus time. The time bases of different trials are synchronized at the time the needle enters the stiff constraint (t = 0, solid blue line). Trajectories are averaged among subjects for each feedback modality, and average trajectories plus/minus standard deviations are shown. The position of the stiff constraint (dashed black line, 100 %) and of the soft tissue phantom surface (dotted black line, 0 %) are shown as well.

Figure 5.13 shows the force registered by the force sensor (solid blue line) and the one applied to the subject (solid green line) versus time. The difference between the

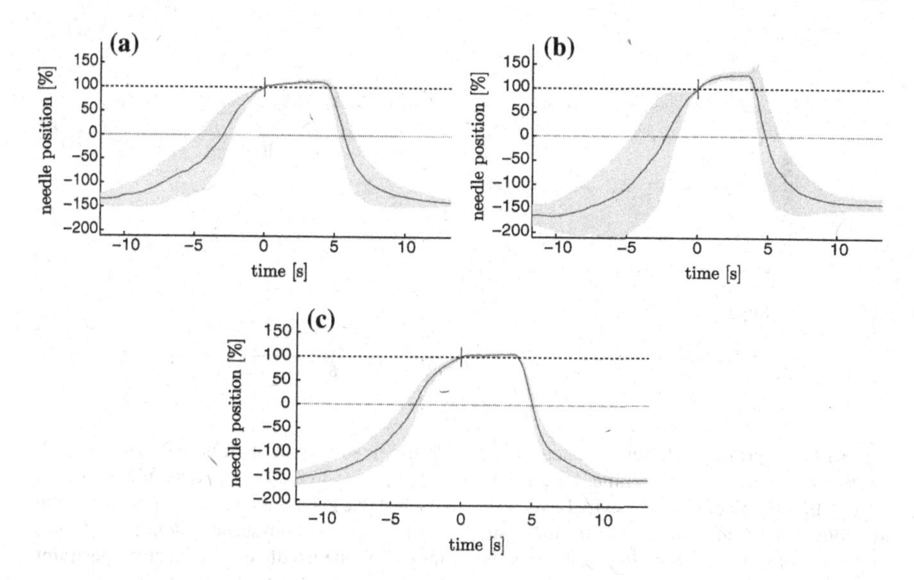

Fig. 5.12 Experiment #2. Average needle trajectory (*solid red line*) and its standard deviation (*orange patch*) are plotted. The position of the stiff constraint (*dashed black line*) and the position of soft tissue phantom surface (*dotted black line*) are shown as well. The *blue line* represents the instant when the needle enters the stiff constraint. **a** Condition F. **b** Condition C. **c** Condition EF (color figure online)

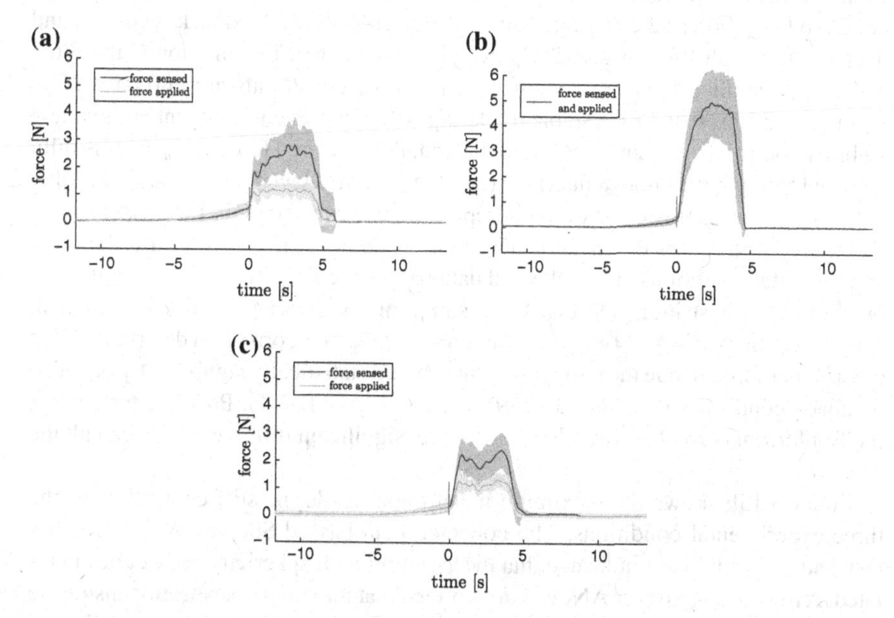

Fig. 5.13 Experiment #2. Teleoperated needle insertion in soft tissue. Average force sensed at the needle tip (*solid blue line*) and force provided to the subject (*solid green line*), together with their standard deviations (*light patches*). The *red line* represents the instant when the needle enters the stiff constraint. **a** Condition F. **b** Condition C. **c** Condition EF (color figure online)

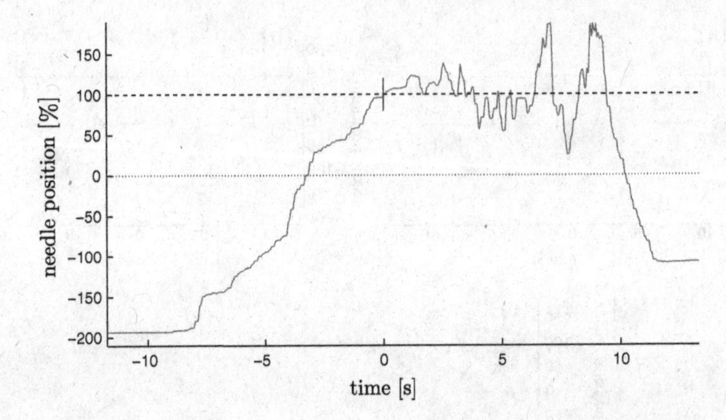

Fig. 5.14 Experiment #2. Teleoperated needle insertion in soft tissue with no passivity control. Position of the needle versus time, for a representative run. Desired force τ_{TLm} is fully rendered through the Omega device (Passivity Layer bypassed). Unstable behavior arises. The position of the stiff constraint (*dashed black line*) and the position of soft tissue phantom surface (*dotted black line*) are shown as well. The *blue line* represents the instant when the needle enters the stiff constraint (color figure online)

blue and green line is a measure of loss of transparency. The time bases of different trials are again synchronized at the time the needle enters the stiff constraint ($t = 0$, solid red line). Forces are averaged among subjects for each feedback modality, and average forces plus/minus standard deviations are shown. In condition C the force sensed and applied is the same, since no passivity constraints are enforced. Note that in conditions F and EF a stable rendering of the virtual environment without any stability control would not be possible. Indeed, if the desired force $\tau_{TLm} = \tau_s$ is fully actuated through the Omega interface (i.e., the passivity layer is bypassed), unstable behavior arises, as it is clear from the representative run shown in Fig. 5.14.

Figure 5.15a shows the mean penetration inside the stiff constraint for the three experimental conditions. The collected data passed the Shapiro-Wilk normality test, but Mauchly's test indicated that the assumption of sphericity had been violated. A repeated measures ANOVA with a Greenhouse-Geisser correction determined that mean penetration inside the stiff constraint differed statistically significantly between feedback conditions ($F(1.384, 26.289) = 72.874$, $p < 0.001$). Post hoc tests using the Bonferroni correction revealed statistically significant difference between all the groups.

Figure 5.15b shows the maximum penetration inside the stiff constraint for the three experimental conditions. The collected data passed Shapiro-Wilk normality test, and Mauchly's test indicated that the assumption of sphericity had not been violated. A repeated measures ANOVA determined that maximum penetration inside the

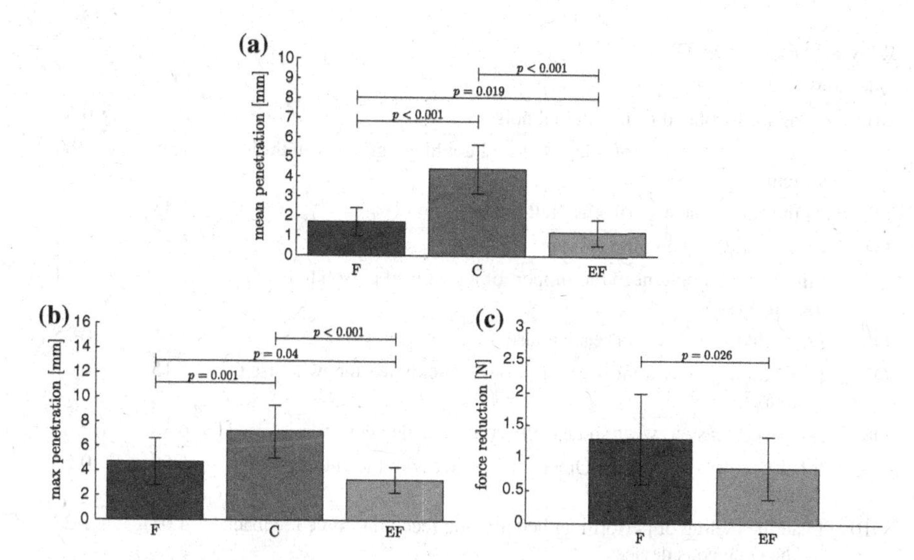

Fig. 5.15 Experiment #2. Teleoperated needle insertion in soft tissue. Mean penetration, maximum penetration and force reduction (mean and SD) for the unaltered method of [10] (F), the cutaneous-only sensory subtraction approach of [13] (C), and the mixed cutaneous-kinesthetic method (EF) are shown. A null value of these metrics indicates the best performance. **a** Mean penetration inside the stiff constraint. **b** Maximum penetration inside the stiff constraint. **c** Force reduction

stiff constraint differed statistically significantly between feedback conditions ($F(2, 38) = 26.128$, $p < 0.001$). Post hoc tests using the Bonferroni correction revealed statistically significant difference between all the groups.

Figure 5.15c shows the average force reduction at the master side, due to passivity constraints, for experimental conditions F and EF. We did not consider data from feedback conditions C since it was not subject to any force reduction. The collected data passed Shapiro-Wilk normality test. A paired-samples t-test determined that the average force reduction at the master side differed statistically significantly between feedback conditions ($t(19) = 2.414$, $p = 0.026$).

No difference between the conditions was observed in terms of task completion time.

In addition to the quantitative evaluation discussed above, we also measured users' experience. Immediately after the experiment, participants were asked to fill in a 11-item questionnaire using bipolar Likert-type seven-point scales. It contained a set of assertions, where a score of 7 was described as "completely agree" and a score of 1 as "completely disagree" with the assertion. The evaluation of each question is reported in Table 5.4.

Table 5.4 Experiment #2

Questions		Mean	σ
Q1	I was well-isolated from external noises	6.40	0.60
Q2	I needed to learn a lot of things before I could get going with this system	1.95	1.00
Q3	At the end of the experiment I felt tired	1.45	0.51
Q4	I felt confident using the system	5.85	0.99
Q5	I think that I would need the support of a technical person to be able to use this system	2.35	1.04
Q6	I thought the system was easy to use	5.65	0.75
Q7	I would imagine that most people would quickly learn how to use this system	6.15	0.67
Q8	It has been easy to wear and use the cutaneous devices	6.40	0.68
Q9	It has been easy to use the Omega 3 together with the cutaneous devices	6.50	0.51
Q10	I had the feeling of performing better while receiving force feedback by the cutaneous devices	4.05	1.27
Q11	I felt hampered by the cutaneous device	1.50	0.76

Users' experience evaluation. Participants rated these statements, presented in random order, using a 7-point Likert scale (1 = completely disagree, 7 = completely agree). Means and standard deviations are reported

5.6 Discussion

Two experiments have been carried out. The first one evaluated the perceived stiffness of a virtual environment, employing the unaltered algorithm in [10] (condition F) and our cutaneous-kinesthetic approach (condition EF). Results are reported in Sect. 5.5.1.2 and Fig. 5.9. The stiffness perceived during repetitions with condition EF was closer to the ideal stiffness than that registered under condition F. The proposed cutaneous-kinesthetic approach was thus more effective in rendering the virtual environment than the unaltered algorithm of [10]. Moreover, since the two feedback conditions share the same underlying passivity controller, they guarantee the same stability properties.

The second experiment evaluated the performance of a 1-DoF teleoperation experiment of needle insertion in soft tissue employing the unaltered algorithm of [10] (condition F), the cutaneous-only sensory subtraction presented in Part I (condition C), and our mixed approach (condition EF). Results are reported in Sect. 5.5.2.2 and Fig. 5.14. The proposed mixed algorithm outperformed the other two feedback conditions in all the metrics considered. As expected, the cutaneous-only sensory subtraction approach performed the worst. However, even under condition C, all the subjects were able to perceive the presence of the stiff constraint and stop the motion of the hand right after the penetration. No difference between the conditions was observed in terms of task completion time. We may read this result by saying that the subjects became equally confident with all the feedback modalities proposed.

Regarding users' experience, subjects felt confident with the system and not hampered by the cutaneous devices. Although results prove differently, subjects did not have the feeling of performing better while receiving additional force feedback from the cutaneous devices.

From these results we can conclude that our methods improved the transparency and performance of the considered teleoperation system with respect to the unaltered algorithm of [10]. The cutaneous-only sensory subtraction approach performed worse than the other two feedback conditions, but still provided a reasonable awareness about the presence of the stiff constraint, as extensively discussed in Part I.

5.7 Conclusions

In this chapter we presented a novel control method to improve transparency of passive teleoperation systems with force reflection, which is based on complementing haptic feedback with additional cutaneous force when a reduction of the force provided by grounded interfaces is required to satisfy stability constraints. The viability of this approach is demonstrated via one experiment of perceived stiffness and one experiment of teleoperated needle insertion in soft tissue. Results showed improved performance with respect to common control techniques not using cutaneous compensation.

The method is rather general and applicable to a wide range of teleoperation systems, provided that each scenario is characterized using perceptual considerations by a suitable mapping function. This approach can be considered as an improvement of the sensory subtraction approach presented in Part I. However, in some scenarios where the safety of the system is paramount, cutaneous only approaches are still preferable, since decoupling actuation and sensing still guarantees a higher degree of safety.

Work is in progress to evaluate the proposed control algorithm in more challenging teleoperation scenarios (e.g., 3-D needle insertion, peg-in-hole tasks). The validation of the proposed approach on top of other energy-based control strategies, as well as the design of ad-hoc controllers for optimal exploitation of cutaneous feedback, are the subject of current research. Moreover, we plan to compare our the proposed method with different feedback techniques, e.g., sensory substitution through visual, vibrotactile or auditory feedback.

References

1. C. Pacchierotti, A. Tirmizi, D. Prattichizzo, Improving transparency in teleoperation by means of cutaneous tactile force feedback. ACM Trans. Appl. Percept. **11**(1), 4:1–4:16 (2014)
2. C. Pacchierotti, A. Tirmizi, G. Bianchini, D. Prattichizzo, Enhancing the performance of passive teleoperation systems via cutaneous feedback. IEEE Trans. Haptics. In Press, (2015)
3. S. Salcudean, Control for teleoperation and haptic interfaces, *Control problems in robotics and automation* (Springer, London, 1998), pp. 51–66

4. P.F. Hokayem, M.W. Spong, Bilateral teleoperation: an historical survey. Automatica **42**(12), 2035–2057 (2006)
5. A.J. van der Schaft, *L2-Gain and Passivity Techniques in Nonlinear Control* (Springer, London, 2000)
6. G. Niemeyer, J.J.E. Slotine, Telemanipulation with time delays. Int. J. Robot. Res. **23**(9), 873–890 (2004)
7. J. Ryu, D. Kwon, B. Hannaford, Stable teleoperation with time-domain passivity control. IEEE Trans. Robot. Autom. **20**(2), 365–373 (2004)
8. J.P. Kim, J. Ryu, Robustly stable haptic interaction control using an energy-bounding algorithm. Int. J. Robot. Res. **29**(6), 666–679 (2010)
9. D. Lee, K. Huang, Passive-set-position-modulation framework for interactive robotic systems. IEEE Trans. Robot. **26**(2), 354–369 (2010)
10. M. Franken, S. Stramigioli, S. Misra, C. Secchi, A. Macchelli, Bilateral telemanipulation with time delays: a two-layer approach combining passivity and transparency. IEEE Trans. Robot. **27**(4), 741–756 (2011)
11. R.E. Schoonmaker, C.G.L. Cao, Vibrotactile force feedback system for minimally invasive surgical procedures. Proc. IEEE Int. Conf. Syst. Man Cybern. **3**, 2464–2469 (2006)
12. M. Kitagawa, D. Dokko, A.M. Okamura, D.D. Yuh, Effect of sensory substitution on suture-manipulation forces for robotic surgical systems. J. Thorac. Cardiovasc. Surg. **129**(1), 151–158 (2005)
13. D. Prattichizzo, C. Pacchierotti, G. Rosati, Cutaneous force feedback as a sensory subtraction technique in haptics. IEEE Trans. Haptics **5**(4), 289–300 (2012)
14. C. Pacchierotti, F. Chinello, M. Malvezzi, L. Meli, D. Prattichizzo, Two finger grasping simulation with cutaneous and kinesthetic force feedback, in *Haptics: Perception, Devices, Mobility, and Communication* (2012), pp. 373–382
15. C. Pacchierotti, F. Chinello, D. Prattichizzo, Cutaneous device for teleoperated needle insertion, in *Proceedings 4th IEEE RAS EMBS International Conference on Biomedical Robotics and Biomechatronics (BioRob)* (2012), pp. 32–37
16. C. Pacchierotti, D. Prattichizzo, K.J. Kuchenbecker, Cutaneous feedback of fingertip deformation and vibration for palpation in robotic surgery, IEEE Trans. Biomed. Eng. (2015), In Press
17. C. Pacchierotti, L. Meli, F. Chinello, M. Malvezzi, D. Prattichizzo, Cutaneous haptic feedback to ensure the stability of robotic teleoperation systems, Int. J. Robot. Res. (2015), http://ijr.sagepub.com/content/early/2015/10/15/0278364915603135.abstract (in press)
18. A. Tirmizi, C. Pacchierotti, D. Prattichizzo, On the role of cutaneous force in teleoperation: subtracting kinesthesia from complete haptic feedback, in *Proceedings of IEEE World Haptics Conference* (2013), pp. 371–376
19. C. Pacchierotti, A. Tirmizi, G. Bianchini, D. Prattichizzo, Improving transparency in passive teleoperation by combining cutaneous and kinesthetic force feedback, in *Proceedings of IEEE/RSJ International Symposium Intelligent Robots and Systems* (2013)
20. K. Minamizawa, S. Fukamachi, H. Kajimoto, N. Kawakami, S. Tachi, Gravity grabber: wearable haptic display to present virtual mass sensation, in *Proceedings of ACM Special Interest Group on Computer Graphics and Interactive Techniques Conference* (2007), 8–es
21. D. Prattichizzo, F. Chinello, C. Pacchierotti, M. Malvezzi, Towards wearability in fingertip haptics: a 3-dof wearable device for cutaneous force feedback. IEEE Trans. Haptics **6**(4), 506–516 (2013)
22. H.L. Stein, Ultrahigh molecular weight polyethylenes(uhmwpe). ASM Int. Eng. Plast. Eng. Mater. Handb. **2**, 167–171 (1988)
23. C.B. Zilles, J.K. Salisbury, A constraint-based god-object method for haptic display. Proc. IEEE/RSJ Int. Conf. Intell. Robots Syst. **3**, 146–151 (1995)
24. G.B. Limentani, M.C. Ringo, F. Ye, M.L. Bergquist, E.O. McSorley, Beyond the t-test: statistical equivalence testing. Anal. Chem. **77**(11), 221–226 (2005)
25. C. Lauzon, B. Caffo, Easy multiplicity control in equivalence testing using two one-sided tests. Am. Stat. **63**(2), 147–154 (2009)

Chapter 6
Cutaneous and Kinesthetic Cues for Enhanced Navigation Feedback in Teleoperation

Abstract In all the previous chapters, haptic feedback was used to render some physical property of the environment, such as its stiffness or resistance to needle penetration. However, this is not the only way the user can benefit from haptic stimuli in robotic teleoperation. There is in fact a growing interest in haptics for guidance, wherein haptic stimuli are used to provide navigation information to the human operator. In this chapter we address this challenge and propose a novel teleop eration system for steering flexible needles. It enables clinicians to directly maneuver the surgical tool while providing them with navigation cues through a combination of kinesthetic and vibrotactile feedback. An ultrasound-guided tracking algorithm tracks in real-time the needle tip and estimates its reachable region. A steering algorithm then computes ideal position and orientation of the needle to always keep the target in its reachable region. The master system uses this information to provide the clinician with haptic feedback about the needle's ideal position and orientation. This information is provided as a combination of kinesthetic and vibrotactile force. Twenty participants carried out an experiment of teleoperated needle insertion into a soft-tissue phantom, considering four different experimental conditions. Participants were provided with either mixed kinesthetic-vibrotactile feedback or mixed kinesthetic-visual feedback. Moreover, we considered two different ways of computing ideal position and orientation of the needle: with or without set-points. Vibrotactile feedback was found more effective than visual feedback in conveying navigation cues.

6.1 Introduction

Advances in teleoperation and robot-assisted surgery have had a large impact in the medical field. Teleoperated robotic surgical systems can greatly improve the accuracy and safety of surgical procedures. They can filter out high-frequency signals and surgical tremor [1], or scale down clinicians' movements to enhance their

---

This chapter is reprinted with kind permission from IEEE, originally published in [2].

C. Pacchierotti, *Cutaneous Haptic Feedback in Robotic Teleoperation*,
Springer Series on Touch and Haptic Systems, DOI 10.1007/978-3-319-25457-9_6

accuracy [3]. Moreover, they may also enable expert clinicians to train or assist other colleagues from a distance, or even directly enable operations from a remote location [4]. Teleoperated robotic systems also improve the ergonomics of the operating theater, since the master interface can be always positioned in a way convenient for the clinician to control [5].

Needle insertion in soft-tissue is a minimally invasive surgical (MIS) procedure used for diagnostic and therapeutic purposes, and it is one of the many surgical procedures that may greatly benefit from the use of teleoperated robotic systems [6]. Inaccurate placement of the needle tip may, in fact, result in misdiagnosis or unsuccessful treatment during, for instance, biopsies or brachytherapies [6, 7]. Hence, researchers have been constantly trying to develop new techniques and systems able to enhance the accuracy of this type of needle insertion. Flexible needles are one of these technological advancements, introduced to provide enhanced steering capabilities [6]. Several control algorithms have been developed for maneuvering flexible needles in two- and three-dimensional spaces. DiMaio and Salcudean presented a path planning and control algorithm which related needle motion at the base (outside the soft-tissue phantom) to the tip motion inside the tissue [8]. Glozman and Shoham [9] and Neubach and Shoham [10] developed an image-guided closed-loop control algorithm for flexible needle steering employing fluoroscopic and ultrasound images, respectively. The needle was modeled as a beam supported by virtual springs, and the forward and inverse kinematics of the needle, used for the two-dimensional (2D) path planning, was reported. Duindam et al. developed a model to describe three-dimensional (3D) deflection of bevel-tipped flexible needles for path planning purposes [11], and Hauser et al. developed a 3D feedback controller to steer needles along a helical path [12]. However, results from both Duindam et al. and Hauser et al. were based solely on simulations, and no experiments in real scenarios were performed. More recently, Abayazid et al. presented a 2D ultrasound image-guided steering algorithm [13] and a 3D needle steering controller for bevel-tipped flexible needles [14], where they used Fiber Bragg Grating sensors to reconstruct the needle shape in real-time.

However, for reasons of safety and acceptance by the medical community, it is often necessary to disregard autonomous approaches and focus more on techniques enabling clinicians to directly control the motion of the medical tools [15–17]. In such a case, the clinician needs to observe, from the master side, the environment the needle is interacting with. This is possible through different types of information that flow from the remote scenario to the human operator. They are usually a combination of visual and haptic stimuli. Visual feedback is already employed in commercial robotic surgery systems (e.g., the da Vinci Si Surgical System) while it is not common to find commercially-available devices providing haptic feedback.

However, as extensively discussed in Chap. 1, force feedback is widely considered to be a valuable navigation tool during teleoperated surgical procedures [3, 18–28].

Wagner et al. [29], for example, examined the effect of haptic force feedback on a blunt dissection task and showed that system performance improved up to 150 % in comparison with providing no force feedback, while also decreasing the number of tissue damaging errors by over a factor of 3. Other studies have linked the lack of significant haptic feedback to increased intraoperative injury in minimally invasive surgery operations [30] and endoscopic surgical operations [31].

Moreover, haptic feedback can be also employed to *augment* the operating environment, providing additional valuable information to the clinician, such as navigation cues. For example, Nakao et al. [32] presented a haptic navigation method that allows clinicians to avoid collision with forbidden regions during surgery. It employs kinesthetic feedback through a 2D master manipulator. More recently, Ren et al. [33] implemented dynamic 3D virtual constraints with haptic and visual feedback during minimally invasive beating-heart procedures.

In addition to these approaches, which mostly involve kinesthetic force feedback, there is also a growing interest in vibrotactile feedback. As discussed in Sect. 1.2.1.2, one of the most notable work employing vibrotactile feedback in robot-assisted surgery has been presented by McMahan et al. [34]. The authors developed a sensing and actuating device for the da Vinci Surgical System able to provide auditory and vibrotactile feedback of tool contact accelerations. Also in Chap. 4 we presented a novel cutaneous feedback system for the da Vinci surgical robot designed to provide planar fingertip deformation and vibration cues to the surgeon.

This chapter presents an innovative teleoperation system for steering flexible needles. It enables clinicians to directly maneuver the surgical tool while providing them with navigation cues through a combination of kinesthetic and vibrotactile feedback. The ultrasound-guided tracking algorithm of Vrooijink et al. [35] tracks in real-time the needle tip and estimates its reachable region. The steering algorithm of Abayazid et al. [14] then computes ideal position and orientation of the needle to always keep the target in its reachable region. The master system uses this information to provide the clinician with haptic feedback about the needle's ideal position and orientation. This information is provided as a combination of kinesthetic and vibrotactile force. A picture of the teleoperation system is reported in Fig. 6.1. Moreover, Fig. 6.2 shows how the master and slave systems are inter-connected.

The main difference between the approach presented by Abayazid et al. and the one presented here is the role of the human operator. In the work of Abayazid et al., the controller has full control on the motion of the slave robot, applying the computed ideal position and orientation directly to the needle. No human is involved in the control loop. However, as mentioned earlier, for reasons of safety and acceptance by the medical community, autonomous robotic control is not desirable [15, 16]. For this reason, in our work, the controller evaluates the ideal position and orientation of the needle but does not directly control the needle's motion. Ideal position and orientation are provided to the master interface, which present them to the clinician, who, in turn, commands the slave robot and steers the needle toward its target point. The clinician has thus *full* control on the motion of the needle, and haptic feedback provides the necessary guiding information. The complexity of the flexible needle kinematics and medical scenario make haptic feedback a valuable support tool for

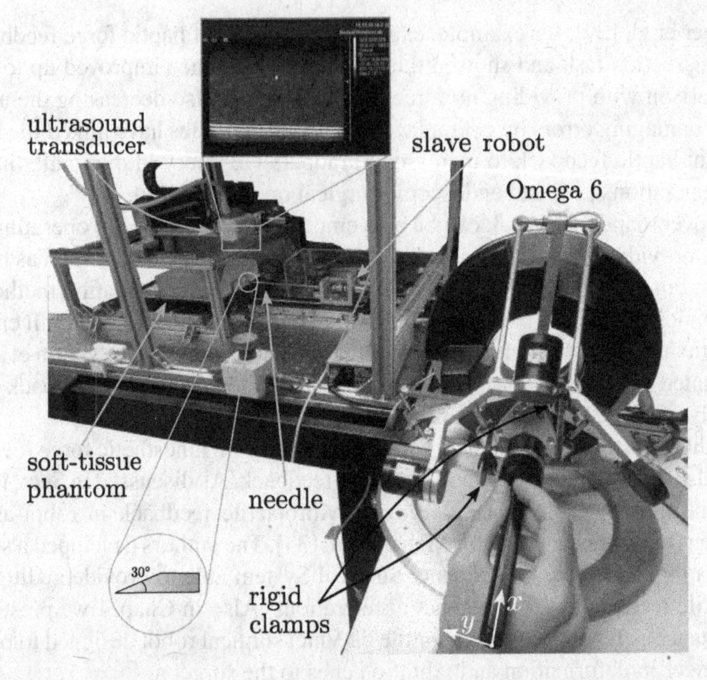

Fig. 6.1 Teleoperation system. Through the Omega 6 haptic device the clinician controls the motion of the slave robot and, thus, the needle. The haptic interface also provides the clinician with navigation cues about the ideal position and orientation of the needle tip, evaluated by the steering algorithm. The needle tip is polished to a bevel angle of 30°

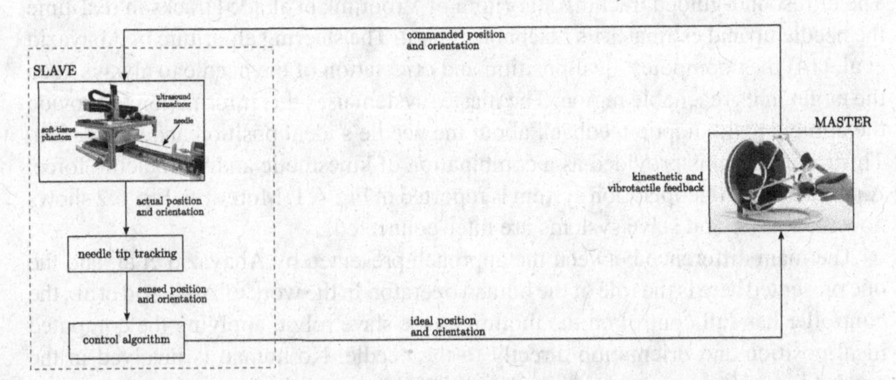

Fig. 6.2 Teleoperation system overview. The ultrasound-guided steering algorithm, described in Sect. 6.3.1, computes the ideal position and orientation of the needle. The haptic interface provides this information to the clinician through a mix of kinesthetic and vibrotactile forces, as described in Sect. 6.3.2. The clinician then controls the motion of the slave robot from the master interface

guidance. Finally, to the best of our knowledge, no commercially-available surgical system provide such a rich pattern of information through the haptic channel.

Section 6.2 presents the teleoperation system, while Sect. 6.3 describes how the navigation stimuli are computed. Sections 6.4 and 6.5 present and discuss an experiment of teleoperated needle insertion in soft tissue, respectively. Finally, Sect. 6.6 addresses concluding remarks and perspectives of the work.

6.2 A Robotic Teleoperation System for Needle Insertion

The slave system consists of a bevel-tipped nitinol needle mounted on a two-degree-of-freedom robotic device. The robot allows the needle to move along the direction of insertion and rotate about its axis (see Fig. 6.3). Moreover, an ultrasound-guided tracking system is used to determine needle tip position during the insertion. The needle's tracking is guaranteed by a 18 MHz ultrasound transducer mounted on a three DoF Cartesian robot, which follows the needle tip during the insertion. The transducer is connected to a Siemens Acuson S2000 ultrasound machine (Siemens AG, Germany). The steering and tracking algorithms are summarized in Sect. 6.3.1.

The master system consists of the single-contact grounded haptic interface Omega 6 (Force Dimension, CH), shown in Fig. 6.4. Two rigid clamps prevent the wrist of the haptic device from moving. The actuators then block two additional DoF, resulting in a haptic interface with 2 DoF, one active (translation in the x direction) and one passive (rotation of the pen-shaped end-effector about the x-axis). The master interface allows the clinician to steer the needle and provides her with navigation

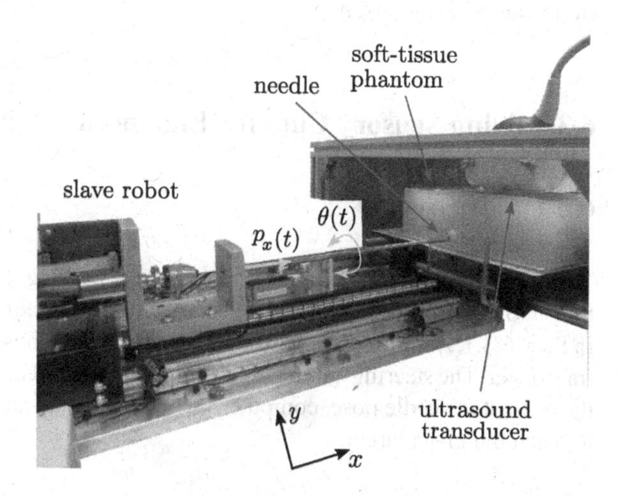

Fig. 6.3 Slave system. The two-degree-of-freedom robotic device steers the flexible needle according to the commanded position $p_x(t)$ and commanded orientation $\theta(t)$, provided by the master device. The ultrasound transducer tracks the needle during the insertion

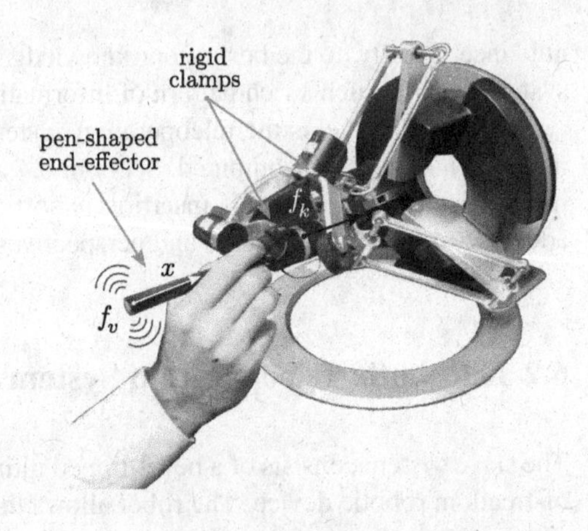

Fig. 6.4 Kinesthetic-vibrotactile feedback. The Omega 6 haptic interface enables the operator to directly steer the needle while being provided with kinesthetic force $\mathbf{f_k}$ and vibrotactile force $\mathbf{f_v}$ about needle's ideal position and orientation, respectively. The motion of the haptic device is constrained along its x-axis. This feedback condition provides no visual information

cues through kinesthetic and vibrotactile feedback. The haptic rendering algorithm is detailed in Sect. 6.3.2.

Communication between the slave and the master systems is set up through a User Datagram Protocol over IP (UDP/IP) socket connection on an Ethernet Local Area Network (LAN). The stability of the teleoperation system is guaranteed by the passivity-based approach presented by Franken et al. [36]. The control algorithm is able to guarantee stable behavior of bilateral telemanipulation systems in the presence of time-varying destabilizing factors, such as stiff control settings, relaxed user grasps, and/or communication delays. The control architecture is split into two separate layers. The hierarchical top layer, named *Transparency Layer*, aims at achieving the desired transparency, while the lower layer, named *Passivity Layer*, ensures the passivity of the system. Further information about this passivity-based control algorithm can be found in Chap. 5.

6.3 Mixing Sensory Cues for Enhanced Navigation

6.3.1 Slave System

The slave system is in charge of tracking and steering the needle during its insertion into the soft-tissue phantom. The needle tip tracking algorithm, summarized in Sect. 6.3.1.1, intra-operatively estimates the needle pose through a 2D ultrasound transducer. The steering control algorithm, summarized in Sect. 6.3.1.2, starting from the estimated needle pose, computes the ideal position and orientation of the needle to reach the given target.

6.3.1.1 Needle Tip Tracking Algorithm

The needle tip tracking algorithm is guided by the ultrasound images obtained from a 2D transducer, which is positioned perpendicularly to the direction of the needle insertion (see Fig. 6.3). A closed-loop control system is in charge of moving the transducer along the needle path so that the tip is always in its field-of-view. The controller is based on a proportional-derivative algorithm, and it aims at minimizing the error between the transducer scanning velocity and the needle insertion velocity at its tip, which is obtained from the slave robot's controller. Furthermore, a Kalman observer minimizes the noise influence on the states of both location and velocity of the needle tip and predicts subsequent states according to the needle tip velocity [37]. Finally, ultrasound images are intra-operatively processed using basic image processing techniques, such as median blur, thresholding, erosion and dilation. This increases the contrast between the tip and the surrounding phantom, preventing false tip detections. After that, the system computes the needle centroid location.

The controller guarantees an accuracy in estimating the needle tip pose up to 0.64 mm and 2.68° for position and orientation, respectively. Further details on the tracking algorithm are presented by Vrooijink et al. [35].

6.3.1.2 3D Needle Steering Control Algorithm

The steering control algorithm intra-operatively estimates the region the needle can reach. Since the needle can be assumed to move along arcs during its insertion [38], this reachable region can be represented by a conical shape (see Fig. 6.5). The direction of each arc depends on the bevel tip orientation, which is controlled by rotating the needle about its insertion axis.

Let us consider a reference frame ψ_{tip}, fixed to the needle tip, that is updated with the changes in the needle pose (position and orientation) estimated by the tracking algorithm of Sect. 6.3.1.1. Target location $\mathbf{p}_{\text{tar}}^{\text{tip}}$, with respect to the needle tip, is thus given by

$$\mathbf{p}_{\text{tar}}^{\text{tip}} = [t_{\text{x}}^{\text{tip}} \; t_{\text{y}}^{\text{tip}} \; t_{\text{z}}^{\text{tip}}]^T , \tag{6.1}$$

where $t_{\text{x}}^{\text{tip}}$, $t_{\text{y}}^{\text{tip}}$, and $t_{\text{z}}^{\text{tip}}$ are the target coordinates along the x_{tip}–, y_{tip}–, and z_{tip}–axis, respectively. Figure 6.5 shows the conical region reachable by the needle (in purple) and the *control* circle (in light blue), which is the circular section of the reachable region intersecting with the centroid of the target. The radius of the control circle r_t is determined by the curvature radius of the needle path r_{c} and by $t_{\text{x}}^{\text{tip}}$, as indicated in Fig. 6.5b. Radius r_{c} has been experimentally measured to be 270 mm [13]. The distance between the target and the center of the control circle is given by

$$d = \sqrt{\left(t_{\text{y}}^{\text{tip}}\right)^2 + \left(t_{\text{z}}^{\text{tip}}\right)^2}. \tag{6.2}$$

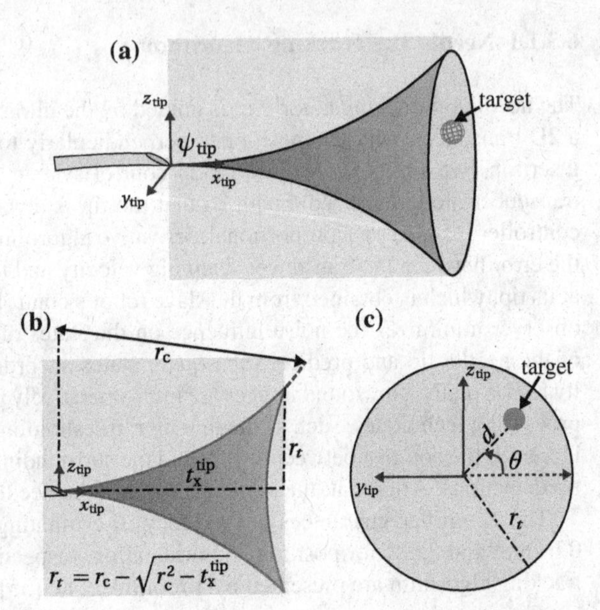

Fig. 6.5 3D needle steering control algorithm. In the *top figure* **a** ψ_{tip} is the frame attached at the tip, while the *blue sphere* represents the target. **b** t_x^{tip} represents the distance between the plane of the control *circle* and the origin of ψ_{tip}, and r_t represents the radius of the control *circle*. Radius r_c is the curvature radius of the *circular* path [38]. **c** d and θ represent the distance of target from the center and the angle of needle rotation, respectively (color figure online)

During the insertion, the radius of the control circle r_t decreases as the needle moves toward the target. This results in the target eventually intersecting with the circumference of the control circle ($d = r_t$). When this happens, the controller computes the new orientation the needle tip should have to keep the target in its reachable region. This ideal orientation $\theta_i(t)$ is given by

$$\theta_i(t) = \tan^{-1}\left(\frac{t_z^{tip}}{t_y^{tip}}\right). \tag{6.3}$$

We assume that no torsion occurs along the needle shaft during the insertion.

On the other hand, the ideal position of the needle during insertion $p_{i,x}(t)$ is computed to maintain a constant insertion speed of 1 mm/s along the x-axis of the slave robot (see Fig. 6.3). Additional details about the steering algorithm can be found in the work by Abayazid et al. [14].

A modified version of this algorithm consists in steering the needle toward the final target through a sequence of preceding points, which we referred to as *set-points*. As the needle tip reaches a set-point, the next one is automatically set to be the goal, and so on, until, eventually, the only point left is the final target. Both approaches, with and without set-points, will be evaluated in Sect. 6.4. Additional details about this modified steering algorithm can be found in the work of Abayazid et al. [39].

6.3.2 Master System

The master system is responsible for both steering the slave robot and displaying navigation cues, i.e., ideal position and orientation of the needle. In order to avoid confusion, and consequent possible errors in the medical intervention, the meaning of such cues must be easy to understand.

In this work we propose to provide the human operator with a combination of kinesthetic and vibrotactile feedback, as detailed in Sect. 6.3.2.1. Conveying information solely through the haptic channel leaves, in fact, the other sensory channels free. For example, an operator teleoperating a needle using this feedback condition may also be provided with additional visual information (e.g., an ultrasound image of the needle). However, force feedback is not the only feedback condition able to convey such information. For this reason we decided to compare our mixed kinesthetic-vibrotactile approach to a different combination of stimuli. This second feedback condition provides the operator with a combination of kinesthetic and visual feedback, as detailed in Sect. 6.3.2.2.

6.3.2.1 Mixing Kinesthetic and Vibrotactile Feedback

Combining multiple haptic stimuli to convey both ideal position and orientation of the needle is an interesting and promising choice. It leaves other sensory channels free and uses a single interface to both steer the slave robot and display navigation cues to the operator. However, we have to face the challenging problem of conveying two stimuli through the haptic sensory channel.

In order to differentiate those stimuli, we propose to provide the operator with

- kinesthetic force $\mathbf{f_k}$ to convey information about the ideal *position* of the needle tip,
- vibrotactile force $\mathbf{f_v}$ to convey information about the ideal *orientation* of the needle tip,

as depicted in Fig. 6.4. Ideal position $p_{i,x}(t) \in \mathbb{R}$ and ideal orientation $\theta_i(t) \in \mathbb{R}$ at time t are computed by the steering algorithm as described in Sect. 6.3.1.2.

Kinesthetic force feedback along the x-axis is controlled by a penalty function based on the distance between the position of the haptic probe $\mathbf{p(t)} = \begin{bmatrix} p_x(t) & p_y(t) & p_z(t) \end{bmatrix}^{\mathrm{T}} \in \mathbb{R}^{3\times1}$ and the current ideal position $p_{i,x}(t)$, while the motion along the y and z axes is blocked:

$$\mathbf{f_k} = \mathbf{K}\,\mathbf{D(t)} - \mathbf{B}\,\dot{\mathbf{D}}\mathbf{(t)}, \tag{6.4}$$

where $\mathbf{B} = 1.5\,\mathbf{I_3}$ Ns/m, $\mathbf{K} = \mathrm{diag}\,[1 \quad 4 \quad 4]$ N/mm, and $\mathbf{D} = \mathbf{p_i(t)} - \mathbf{p(t)}$ is the distance between the ideal position $\mathbf{p_i(t)} = [p_{i,x}(t) \;\; 0.10 \;\; 0.08]^{\mathrm{T}}$ m and the current position of the haptic probe. The motion is thus limited along the x-axis and a kinesthetic force guides the operator toward $p_{i,x}(t)$.

On the other hand, information concerning the orientation of the needle tip is provided through vibrotactile feedback. Vibrations are controlled by a penalty function based on the difference between the ideal orientation $\theta_i(t)$ and the current orientation of the haptic probe $\theta(t) \in \mathbb{R}$:

$$\mathbf{f}_v = \mathbf{A}_1 \, |\theta_i(t) - \theta(t)| \, \operatorname{sgn}(\sin(2\pi f t)), \tag{6.5}$$

where $\mathbf{A}_1 = \frac{3}{\pi}\mathbf{I}_{3\times 1}$ N/rad and

$$f = \begin{cases} 200\,\text{Hz} & \text{if } \theta(t) - \theta_i(t) \geq 0, \\ 150\,\text{Hz} & \text{if } \theta(t) - \theta_i(t) < 0. \end{cases}$$

Vibrations thus provide information about the ideal orientation $\theta_i(t)$, indicating in which direction and how much the clinician should rotate the pen-shaped haptic probe. Frequency f indicates in which direction the clinician should rotate the pen-shaped haptic probe: clockwise for $f = 200$ Hz and counter-clockwise for $f = 150$ Hz. Frequency values are chosen to maximally stimulate the Pacinian corpuscle receptors [40], be easy to distinguish [41] and fit the master device specifications. On the other hand, the amplitude of these vibrations indicates how much the clinician should rotate the haptic probe: no vibrations indicated the best performance. Amplitude scaling matrix $\mathbf{A}_1$ is chosen to maximize the just-noticeable difference [42] for the error $|\theta_i(t) - \theta(t)|$ and fit the master device specifications.

The total force provided to the operator through the Omega 6 haptic interface is then evaluated by combining Eqs. (6.4) and (6.5),

$$\mathbf{f}_{t,1} = \mathbf{f}_k + \mathbf{f}_v. \tag{6.6}$$

The operator is asked to keep the magnitude of $\mathbf{f}_{t,1}$ as small as possible, since a null value of this force denotes the least error.

6.3.2.2 Mixing Kinesthetic and Visual Feedback

In the above mentioned approach all the information is conveyed through the haptic interface: kinesthetic force for position and vibrotactile stimuli for orientation. In order to better evaluate the effectiveness of our mixed kinesthetic-vibrotactile approach, we compared it to a different combination of stimuli, employing a popular visual feedback technique [23]. In this alternative feedback condition we provide the operator with

- kinesthetic force $\mathbf{f}_k$ to convey information about the ideal *position* of the needle tip (as in the previous kinesthetic-vibrotactile approach),
- visual feedback to convey information about the ideal *orientation* of the needle tip,

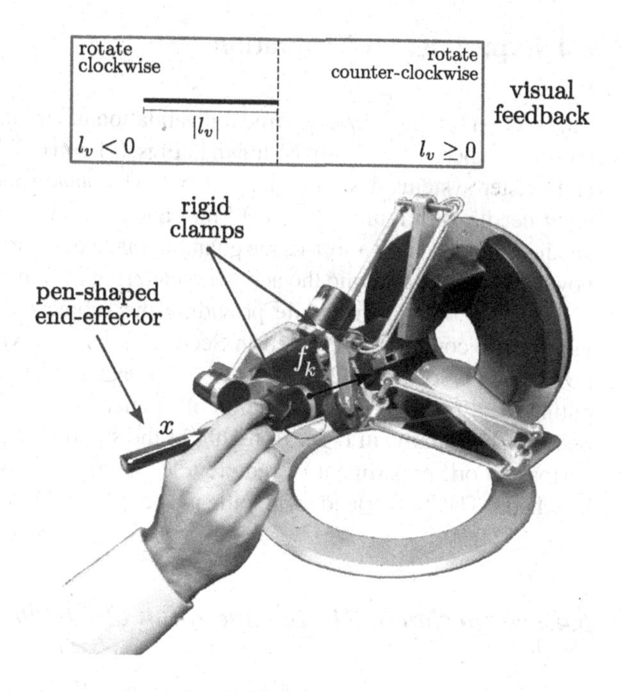

Fig. 6.6 Kinesthetic-visual feedback. The Omega 6 haptic interface enables the operator to directly steer the needle while being provided with kinesthetic force $\mathbf{f_k}$ and visual feedback about needle's ideal position and orientation, respectively

as depicted in Fig. 6.6. Ideal position $p_{i,x}(t) \in \mathbb{R}$ and ideal orientation $\theta_i(t) \in \mathbb{R}$ at time t are again computed by the steering algorithm as described in Sect. 6.3.1.2.

Kinesthetic force feedback is computed as in the kinesthetic-vibrotactile condition (see Eq. 6.4). The motion is again limited along the x-axis and a kinesthetic force guides the operator toward $p_{i,x}(t)$. Moreover, since this time kinesthetic force is the only force applied to the operator, we can define the total force provided through the Omega 6 simply as

$$\mathbf{f}_{t,2} = \mathbf{f_k}. \tag{6.7}$$

On the other hand, information concerning the orientation of the needle tip is now provided through visual feedback. A black horizontal bar shows on the screen the difference between the ideal orientation $\theta_i(t)$ and the current orientation of the haptic probe $\theta(t)$. Its height is fixed to 5 mm and its width varies as

$$l_v = A_2 \, |\theta_i(t) - \theta(t)|,$$

where $A_2 = \frac{10}{\pi}$ cm/rad. The operator is asked to keep l_v as small as possible, since a null value of the bar width denotes the least error. If $l_v < 0$ the bar grows on the left, and the operator thus needs to rotate the pen-shaped end-effector clockwise (as in Fig. 6.6). Otherwise, if $l_v \geq 0$, the bar grows on the right, and the operator is required to rotate the end-effector counter-clockwise. Amplitude scaling matrix A_2 is chosen to provide good sensitivity to the error $|\theta_i(t) - \theta(t)|$ and be seen in one glance, without the need of rotating the head.

6.4 Experimental Evaluation

This section presents the experimental validation of the integrated teleoperation system. The experimental setup is shown in Figs. 6.1 and 6.7. It is composed of the slave and master systems described in Sect. 6.2. The slave robot steers a flexible nitinol alloy needle with a diameter of 0.5 mm and a bevel angle (at the tip) of 30°. The needle is inserted into a soft-tissue phantom made of gelatine mixture, to which silica powder is added to mimic the acoustic scattering of human tissue [14].

We tested our system while providing the subjects with either the kinesthetic-vibrotactile condition presented in Sect. 6.3.2.1 or the kinesthetic-visual condition presented in Sect. 6.3.2.2. Moreover, we considered two different ways of computing ideal position and orientation of the needle: with or without set-points (see Sect. 6.3.1). Finally, in order to highlight the stability properties of the system, we performed one experiment of remote teleoperation: the slave robot was located in Enschede, The Netherlands while the master system was located in Genova, Italy.

6.4.1 Experiment #1: Teleoperation of Flexible Needles

Twenty subjects (12 males, 8 females, age range 23–56 years) took part in the experiment, all of whom were right-handed. Four of them had previous experience with haptic interfaces. None reported any deficiencies in their perception abilities.

The task consisted of steering the needle toward a given target point, located at $\mathbf{o_t} = [85\ -10\ \ 5]^T$ mm with respect to the initial position of the needle (see Fig. 6.3). The control algorithm calculates the ideal position and orientation of the needle tip using either set-points or not, as discussed in Sect. 6.3.1. The haptic interface

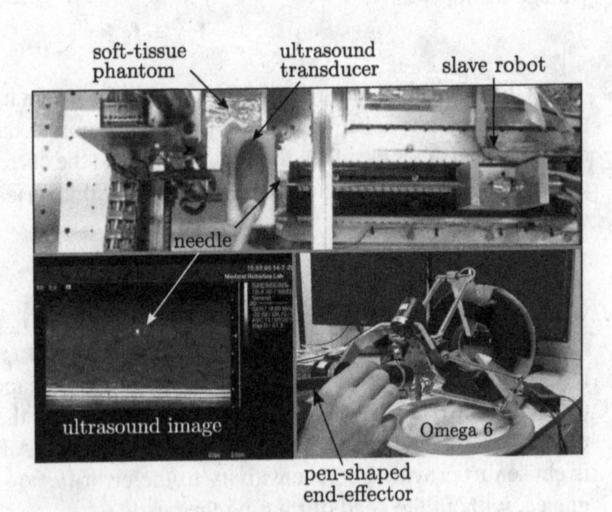

Fig. 6.7 Experimental test. The operator, through the haptic device (*bottom-right*), steers the needle inside the soft-tissue phantom (*top*). The controller, through the ultrasound imaging system (*bottom-left*), tracks the needle and evaluates the ideal position and orientation of the needle tip, either using set-points or not. Navigation cues are provided to the operator via either kinesthetic-vibrotactile or kinesthetic-visual feedback

presents these two pieces of information either via kinesthetic-vibrotactile feedback or kinesthetic-visual feedback, as discussed in Sect. 6.3.2. The operator then steers the needle, relying only on these navigation cues. A short video of the experiment is available as supplemental material at http://extras.springer.com/978-3-319-25455-5 and at http://goo.gl/nzAIuG. Each subject made four randomized trials of the needle steering task, with one repetition for each condition proposed:

- kinesthetic-vibrotactile feedback with ideal position and orientation calculated using set-points (condition VB+S),
- kinesthetic-vibrotactile feedback with ideal position and orientation calculated without using set-points (condition VB),
- kinesthetic-visual feedback with ideal position and orientation calculated using set-points (condition VI+S),
- kinesthetic-visual feedback with ideal position and orientation calculated without using set-points (condition VI),

as summarized in Table 6.1. No visual feedback was provided in conditions VB and VB+S. Moreover, in order to avoid providing undesired auditory cues, subjects were isolated from external noises through a pair of noise-cancelling headphones. Subjects were informed about the procedure before the beginning of the experiment and a 10-min familiarization period was provided to acquaint them with the experimental setup. The mean error in reaching the target point e_t, and the mean errors over time in following the ideal position and orientation signals, e_p and e_o, provided a measure of accuracy. Error e_t is calculated as $\|\mathbf{n_f} - \mathbf{o_t}\|$, where $\mathbf{n_f} \in \mathbb{R}^{3\times1}$ represents needle tip position at the end of the task. Errors on the ideal signals, e_p and e_o, are computed as the mean over time of $\|p_x(t) - p_{i,x}(t)\|$ and $\|\theta(t) - \theta_i(t)\|$, respectively. A null value of these three metrics denotes the best performance. Figure 6.8 shows commanded and ideal orientation for a representative run of the experiment (condition VB).

Figure 6.9a shows targeting error e_t for the four experimental conditions. Comparison of the means among the feedback conditions was tested using a two-way repeated-measures ANOVA. Type of feedback (kinesthetic-vibrotactile and kinesthetic-visual) and way of computing ideal signals (with and without set-points) were considered as within-subject factors. The collected data passed Shapiro–Wilk normality test. The means differed significantly among types of feedback ($F_{1,19} = 9.744$, p = 0.006, a = 0.05) and ways of computing ideal signals ($F_{1,19} = 26.517$, $p < 0.001$, a = 0.05).

Table 6.1 The four experimental conditions tested

Controller	Feedback condition	
	Kinesthetic-vibrotactile	Kinesthetic-visual
Set-point	VB+S	VI+S
No set-point	VB	VI

Ideal position and orientation were either calculated with or without set-points. Information about ideal signals was provided through a mix of kinesthetic force and either visual or vibrotactile cues

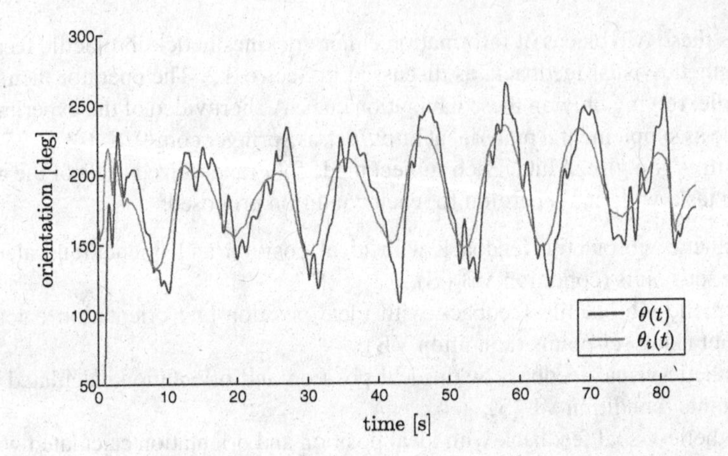

Fig. 6.8 Experimental evaluation (condition VB). Commanded and ideal orientation for a representative run are shown in *blue* and *red*, respectively (color figure online)

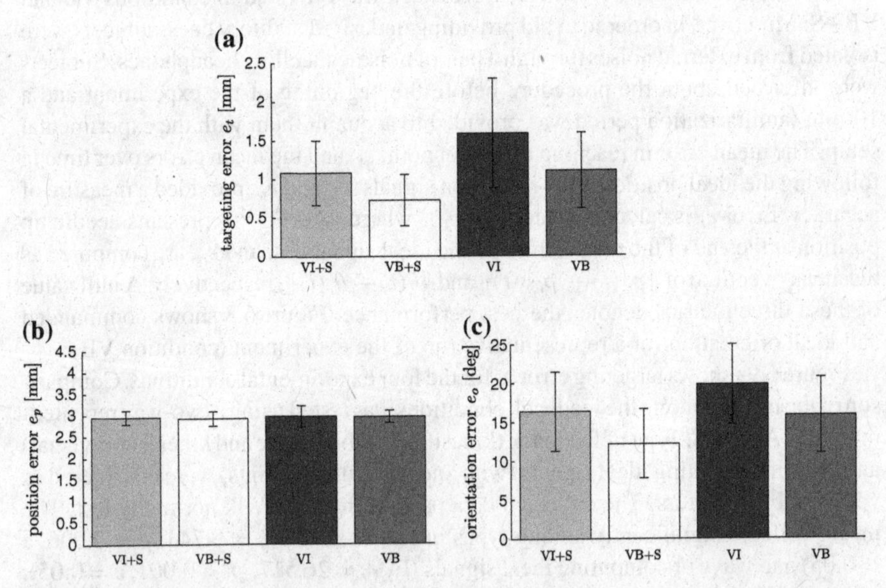

Fig. 6.9 Needle insertion experiment. Targeting error e_t, position error e_p, and orientation error e_o (mean and SD are plotted) for the kinesthetic-visual w/set-points (VI+S), kinesthetic-visual w/o set-points (VI), kinesthetic-vibrotactile w/set-points (VB+S), and kinesthetic-vibrotactile w/o set-points (VI) conditions. Lower values of these metrics indicate higher performances in completing the given task. **a** Targeting error e_t. **b** Position error e_p. **c** Orientation error e_o

Figure 6.9b shows position error e_p for the four experimental conditions. Comparison of the means among the feedback conditions was tested using a two-way repeated-measures ANOVA. Type of feedback (kinesthetic-vibrotactile and kinesthetic-visual) and way of computing ideal signals (with and without set-points) were considered as within-subject factors. The collected data passed Shapiro–Wilk normality test. The means did not differed significantly among types of feedback ($F_{1,19} = 0.314$, $p = 0.582$, $a = 0.05$) and ways of computing ideal signals ($F_{1,19} = 0.366$, $p = 0.553$, $a = 0.05$).

Figure 6.9c shows orientation error e_o for the four experimental conditions. Comparison of the means among the feedback conditions was tested using a two-way repeated-measures ANOVA. Type of feedback (kinesthetic-vibrotactile and kinesthetic-visual) and way of computing ideal signals (with and without set-points) were considered as within-subject factors. The collected data passed Shapiro–Wilk normality test. The means differed significantly among types of feedback ($F_{1,19} = 12.501$, $p = 0.002$, $a = 0.05$) and ways of computing ideal signals ($F_{1,19} = 11.463$, $p = 0.003$, $a = 0.05$).

The experiment lasted 7.1 min on average.

In addition to the quantitative evaluation reported above, we also measured users' experience. Immediately after the experiment, subjects were asked to fill in a 12-item questionnaire using bipolar Likert-type seven-point scales. It contained a set of assertions, where a score of 7 was described as "completely agree" and a score of 1 as "completely disagree" with the assertion. It was divided in three sections considering (i) general statements on the system and perceived performance while using either (ii) kinesthetic-vibrotactile feedback or (iii) kinesthetic-visual feedback. The evaluation of each question is reported in Table 6.2. A Wilcoxon Signed-Rank

Table 6.2 Users' experience evaluation

	Questions		Mean	σ
General	Q1	I was well-isolated from external noises	6.10	0.64
	Q2	I had the feeling of performing better while receiving vibrations from the interface	4.15	1.27
	Q3	I needed to learn a lot of things before I could get going with this system	1.65	0.67
	Q4	At the end of the experiment I felt tired.	2.05	1.32
VB and VB+S	Q5	I felt confident using the system	4.90	0.94
	Q6	I think that I would need the support of a technical person to be able to use this system	2.10	0.85
	Q7	I thought the system was easy to use	5.75	0.72
	Q8	I would imagine that most people would quickly learn how to use this system	5.60	0.94
VI and VI+S	Q9	I felt confident using the system	5.71	1.35
	Q10	I think that I would need the support of a technical person to be able to use this system	1.80	0.52
	Q11	I thought the system was easy to use	6.50	0.61
	Q12	I would imagine that most people would quickly learn how to use this system	6.15	0.81

Participants rated these statements, presented in random order, using a 7-point Likert scale (1 = completely disagree, 7 = completely agree). Means and standard deviations are reported for the kinesthetic-vibrotactile w/set-points (VB+S), kinesthetic-vibrotactile w/o set-points (VI), kinesthetic-visual w/set-points (VI+S), and kinesthetic-visual w/o set-points (VI) conditions

Test was performed to evaluate the statistical significance of the difference between questions proposed for both feedback modalities, i.e., Q5 versus Q9, Q6 versus Q10, Q7 versus Q11, and Q8 versus Q12. The p-values revealed a statistically significant difference between Q7 and Q11 ($Z = -3.274$, $p = 0.001$), and between Q8 and Q12 ($Z = -2.209$, $p = 0.027$).

6.4.2 Experiment #2: Teleoperation in a Remote Scenario

As mentioned in Sect. 6.2, communication between the master and slave systems is set through a UDP/IP socket connection. For this reason, although the experiment presented in Sect. 6.4.1 considers a scenario in which the robotic systems are connected to the same LAN, master and slave could be easily placed in different LANs and then communicate through a common internet connection. The use of internet as a mean of communication in bilateral teleoperation has, in fact, lately gained increasing attention due to its cost-effective and flexible applications [43]. However, this type of digital transmission exchanges data packets through a network characterized by significant variations in time delays. Such a network may also cause unreliable communication due to the loss of packets associated with the considered channel congestion [43, 44]. As a result, bilateral teleoperation performance may severely degrade, and unstable behaviors may arise. However, the passivity-based controller employed in this work is able to guarantee the stability of our teleoperation system even in such cases.

In order to evaluate the safety, stability and performance of our system in the presence of such destabilizing factors, we carried out one additional repetition of the needle insertion experiment. The same setup and protocol presented in Sect. 6.4 and shown in Figs. 6.1 and 6.7 were employed. The only difference was the master and slave systems being connected through a Digital Subscriber Line (DSL) internet network. The master robot was placed in Genova, Italy, while the slave system was in Enschede, The Netherlands. I was the subject (26 years old, male, right-handed). We considered experimental condition VB. A video of the remote experiment is available as supplemental material at http://extras.springer.com/978-3-319-25455-5 and at http://goo.gl/nzAIuG (seek to minute 3:37). Results showed a targeting error $e_t = 0.71$ mm, an error in following the ideal position $e_p = 2.87$ mm, and an average error in following the ideal orientation $e_o = 13.26\,°$. An average round-trip time of 56.3 ms was registered during the needle insertion experiment. Packet loss was negligible.

6.5 Discussion

We tested four experimental conditions, considering two ways of computing ideal stimuli, with and without set-points, and two feedback conditions, kinesthetic-vibrotactile and kinesthetic-visual. The average error in reaching the target point

e_t, and the average errors in following the ideal position and orientation signals, e_p and e_o, provided a measure of accuracy.

Results are reported in Sect. 6.4.1 and Fig. 6.9. Targeting errors e_t and orientation errors e_o were lower for kinesthetic-vibrotactile conditions VB and VB+S with respect to kinesthetic-visual conditions VI and VI+S, respectively. Vibrotactile stimuli were thus more effective in conveying navigation cues than visual ones. This confirms the result found by Ramos and Prattichizzo [45], where vibrotactile feedback showed a significantly lower reaction time with respect to visual feedback. Moreover, conditions VB+S and VI+S, which use set-points, showed better performance than conditions VB and VI. On the other hand, as expected, no difference was found in position errors e_p. All the four experimental conditions, in fact, provide navigation cues about ideal position through kinesthetic force. However, although results show better performance when providing vibrotactile feedback, users considered the visual condition easier and more intuitive. This can be explained by considering that humans are way more used to deal with visual cues with respect to vibrotactile ones, and, therefore, they feel more comfortable with them.

Regarding the comparison with autonomous approaches, participants here showed worse performance with respect to the autonomous controller presented by Abayazid et al. [14], where the steering algorithm controlled directly the slave robot. Abayazid et al. achieved a mean targeting error e_t of 1.3 mm when steering the needle in an autonomous way. However, the targeting accuracy achieved in our work is still sufficient to reach the smallest lesions detectable using state-of-art ultrasound imaging systems (ϕ 2 mm). Moreover, our results outperform MRI-guided biopsies carried out directly by clinician (no robots involved). El Khouli et al. [46], in fact, found a mean 3D biopsy targeting error of 4.4 ± 2.9 mm for biopsies of phantoms, and a mean 3D localization error of 5.7 ± 3.0 mm for breast biopsies performed in patients.

Finally, we tested our system in a remote teleoperation scenario. Results were comparable to the one registered in Sect. 6.4.1 and no unstable behavior arose, assessing the effectiveness of the stability controller employed.

6.6 Conclusions

In this chapter we presented a robotic teleoperation system for needle steering, able to provide the clinician with navigation cues through a mix of kinesthetic and vibroactile force fededback. It is composed of a slave and a master system. The slave is composed of a two DoF robotic device to insert the needle and a three DoF Cartesian robot to control the ultrasound transducer that tracks the needle tip during insertion. The master is an Omega 6 haptic interface, in charge of tracking the position of the human hand while providing the operator with navigation cues. In order to evaluate the performance of the proposed system, twenty participants carried out an experiment of teleoperated needle insertion in a soft-tissue phantom, considering four different experimental conditions. Subjects were provided with either mixed kinesthetic-vibrotactile feedback or mixed kinesthetic-visual feedback. Moreover, we considered two different ways of computing ideal position and orientation of the needle: with or without set-points.

Results showed vibrotactile feedback to be more effective than visual feedback in conveying navigation cues. Moreover, conveying information solely through the haptic channel leaves other sensory channels free. For example, a clinician teleoperating a needle with our system may also be provided with additional visual information, e.g., an ultrasound image of the needle. However, results showed worse performance with respect to autonomous insertions, i.e., where the steering algorithm controls directly the slave robot [14]. Nonetheless, the registered targeting accuracy is still sufficient to reach the smallest lesions detectable using state-of-art ultrasound imaging systems. In order to validate the stability properties of our system, we also carried out an experiment of remote teleoperation, in which the master and slave systems were connected through a common DSL internet connection. Results were comparable to the one registered in the first experiment and no unstable behavior arose.

Work is in progress to evaluate the proposed teleoperation system in different clinically-relevant scenarios. We plan to test the proposed system with different target points, different algorithms to calculate ideal signals, introducing obstacles to avoid, and using biological tissue. We also plan to substitute kinesthetic feedback with stimuli of another sensory modality (e.g., cutaneous, audio), in order to make the system intrinsically passive and safe, with no need of stability control (see Part I). Moreover, in order to analyze relevant within-subject effects, each participant will run more then one trial per feedback condition. Finally, work is in progress to use kinesthetic force to provide clinicians with force feedback regarding the mechanical properties of the tissue being penetrated.

References

1. C. Pacchierotti, M. Abayazid, S. Misra, D. Prattichizzo, Teleoperation of steerable flexible needles by combining kinesthetic and vibratory feedback. IEEE Trans. Haptics **7**(4), 551–556 (2014)
2. A.M. Okamura, Methods for haptic feedback in teleoperated robot-assisted surgery. Ind. Robot: Int. J. **31**(6), 499–508 (2004)
3. D. De Lorenzo, E. De Momi, I. Dyagilev, R. Manganelli, A. Formaglio, D. Prattichizzo, M. Shoham, G. Ferrigno, Force feedback in a piezoelectric linear actuator for neurosurgery. Int. J. Med. Robot. Comput. Assist. Surg. **7**(3), 268–275 (2011)
4. J. Marescaux, J. Leroy, F. Rubino, M. Smith, M. Vix, M. Simone, D. Mutter, Transcontinental robot-assisted remote telesurgery: feasibility and potential applications. Ann. Surg. **235**(4), 487 (2002)
5. R.H. Taylor, D. Stoianovici, Medical robotics in computer-integrated surgery. IEEE Trans. Robot. Autom. **19**(5), 765–781 (2003)
6. N. Abolhassani, R. Patel, M. Moallem, Needle insertion into soft tissue: a survey. Med. Eng. Phys. **29**(4), 413–431 (2007)

7. C. Papalazarou, P. Rongen, et al., Surgical needle reconstruction using small-angle multi-view x-ray, in *Proceeding of the IEEE International Conference on Image Processing (ICIP)* (2010), pp. 4193–4196
8. S.P. DiMaio, S.E. Salcudean, Needle insertion modeling and simulation. IEEE Trans. Robot. Autom. **19**(5), 864–875 (2003)
9. D. Glozman, M. Shoham, Image-guided robotic flexible needle steering. IEEE Trans. Robot. **23**(3), 459–467 (2007)
10. Z. Neubach, M. Shoham, Ultrasound-guided robot for flexible needle steering. IEEE Trans. Biomed. Eng. **57**(4), 799–805 (2010)
11. V. Duindam, R. Alterovitz, S. Sastry, K. Goldberg, Screw-based motion planning for bevel-tip flexible needles in 3D environments with obstacles, in *Proceeding of the IEEE International Conference on Robotics and Automation* (2008), pp. 2483–2488
12. K. Hauser, R. Alterovitz, N. Chentanez, A. M. Okamura, K. Goldberg, Feedback control for steering needles through 3D deformable tissue using helical paths, in *Proceeding of the Robotics: Science and Systems (RSS)*, vol. 37 (2009)
13. M. Abayazid, R.J. Roesthuis, R. Reilink, S. Misra, Integrating deflection models and image feedback for real-time flexible needle steering. IEEE Trans. Robot. **29**(2), 542–553 (2013). ISSN: 1552-3098
14. M. Abayazid, M. Kemp, S. Misra, 3D flexible needle steering in soft-tissue phantoms using Fiber Bragg Grating sensors, in *Proceeding of the IEEE International Conference on Robotics and Automation* (2013), pp. 5823–5829
15. J. Troccaz, Y. Delnondedieu, Semi-active guiding systems in surgery. a two-dof prototype of the passive arm with dynamic constraints (padyc). Mechatronics **6**(4), 399–421 (1996). Mechatronics in Surgery
16. M. Jakopec, F. Rodriguez y Baena, S.J. Harris, P. Gomes, J. Cobb, B.L. Davies, The handson orthopaedic robot "acrobot": early clinical trials of total knee replacement surgery. IEEE Trans. Robot. Autom. **19**(5), 902–911 (2003)
17. J.M. Romano, R.J. Webster, A.M. Okamura, Teleoperation of steerable needles, in *Proceeding of the IEEE International Conference on Robotics and Automation* (2007), pp. 934–939
18. M.J. Massimino, T.B. Sheridan, Teleoperator performance with varying force and visual feedback. Hum. Factors: The J. Hum. Factors Ergon. Soc. **36**(1), 145–157 (1994)
19. S.E. Salcudean, S. Ku, G. Bell, Performance measurement in scaled teleoperation for microsurgery, in *Proceeding of the First Joint Conference on Computer Vision, Virtual Reality and Robotics in Medicine and Medial Robotics and Computer-Assisted Surgery* (1997), pp. 789–798
20. A. Kazi, Operator performance in surgical telemanipulation. Presence: Teleoperators Virtual Environ. **10**(5), 495–510 (2001)
21. C.W. Kennedy, T. Hu, J.P. Desai, A.S. Wechsler, J.Y. Kresh, A novel approach to robotic cardiac surgery using haptics and vision. Cardiovasc. Eng. **2**(1), 15–22 (2002)
22. A. Pillarisetti, M. Pekarev, A.D. Brooks, J.P. Desai, Evaluating the effect of force feedback in cell injection. IEEE Trans. Autom. Sci. Eng. **4**(3), 322–331 (2007)
23. D. Prattichizzo, C. Pacchierotti, G. Rosati, Cutaneous force feedback as a sensory subtraction technique in haptics. IEEE Trans. Haptics **5**(4), 289–300 (2012)
24. C. Pacchierotti, F. Chinello, D. Prattichizzo, Cutaneous device for teleoperated needle insertion, in *Proceeding of the 4th IEEE RAS EMBS International Conference on Biomedical Robotics and Biomechatronics (BioRob)* (2012), pp. 32–37
25. C. Pacchierotti, D. Prattichizzo, K.J. Kuchenbecker, Cutaneous feedback of fingertip deformation and vibration for palpation in robotic surgery. Trans. Biomed. Eng. (2015), In Press
26. C. Pacchierotti, A. Tirmizi, G. Bianchini, D. Prattichizzo, Enhancing the performance of passive teleoperation systems via cutaneous feedback. IEEE Trans. Haptics. In Press (2015)
27. C. Pacchierotti, A. Tirmizi, G. Bianchini, D. Prattichizzo, Improving transparency in passive teleoperation by combining cutaneous and kinesthetic force feedback, in *Proceeding of the IEEE/RSJ International Symposium Intelligent Robots and Systems* (2013)

28. C. Pacchierotti, M. Abayazid, S. Misra, D. Prattichizzo, Steering of flexible needles combining kinesthetic and vibratory force feedback, in *Proceedings of IEEE/RSJ International Conference Intelligent Robots and Systems (IROS)* (Chicago, USA, 2014), pp. 1202–1207
29. C.R. Wagner, N. Stylopoulos, R.D. Howe, The role of force feedback in surgery: analysis of blunt dissection, in *Proceeding of the Symposium of Haptic Interfaces for Virtual Environment and Teleoperator Systems* (2002), pp. 68–74
30. A.M. Okamura, Haptic feedback in robot-assisted minimally invasive surgery. Curr. Opin. Urol. **19**(1), 102 (2009)
31. M. Hashizume, M. Shimada, M. Tomikawa, Y. Ikeda, I. Takahashi, R. Abe, F. Koga, N. Gotoh, K. Konishi, S. Maehara, K. Sugimachi, Early experiences of endoscopic procedures in general surgery assisted by a computer-enhanced surgical system. Surg. Endosc. **16**(8), 1187–1191 (2002)
32. M. Nakao, K. Imanishi, T. Kuroda, H. Oyama, Practical haptic navigation with clickable 3D region input interface for supporting master-slave type robotic surgery. Stud. Health Technol. Inf. pp. 265–271 (2004)
33. J. Ren, R.V. Patel, K.A. McIsaac, G. Guiraudon, T.M. Peters, Dynamic 3-D virtual fixtures for minimally invasive beating heart procedures. IEEE Trans. Med. Imaging **27**(8), 1061–1070 (2008)
34. W. McMahan, J. Gewirtz, D. Standish, P. Martin, J.A. Kunkel, M. Lilavois, A. Wedmid, D.I. Lee, K.J. Kuchenbecker, Tool contact acceleration feedback for telerobotic surgery. IEEE Trans. Haptics **4**(3), 210–220 (2011)
35. G.J. Vrooijink, M. Abayazid, S. Misra, Real-time three-dimensional flexible needle tracking using two-dimensional ultrasound, in *Proceeding of the IEEE International Conference on Robotics and Automation* (2013), pp. 1680–1685
36. M. Franken, S. Stramigioli, S. Misra, C. Secchi, A. Macchelli, Bilateral telemanipulation with time delays: a two-layer approach combining passivity and transparency. IEEE Trans. Robot. **27**(4), 741–756 (2011)
37. Y. Bar-Shalom, X.R. Li, T. Kirubarajan, *Estimation with Applications to Tracking and Navigation: Theory Algorithms and Software* (Wiley, New York, 2001)
38. R.J. Webster, J.S. Kim, N.J. Cowan, G.S. Chirikjian, A.M. Okamura, Nonholonomic modeling of needle steering. Int. J. Robot. Res. **25**(5–6), 509–525 (2006)
39. M. Abayazid, G.J. Vrooijink, S. Patil, R. Alterovitz, S. Misra, Experimental evaluation of ultrasound-guided 3D needle steering in biological tissue. Int. J. Comput. Assist. Radiol. Surg. 1–9 (2014)
40. R.W. Cholewiak, A.A. Collins, Sensory and physiological bases of touch (1991)
41. K.A. Kaczmarek, J.G. Webster, P. Bach-y-Rita, W.J. Tompkins, Electrotactile and vibrotactile displays for sensory substitution systems. IEEE Trans. Biomed. Eng. **38**(1), 1–16 (1991)
42. H. Pongrac, Vibrotactile perception: examining the coding of vibrations and the just noticeable difference under various conditions. Multimed. Syst. **13**(4), 297–307 (2008)
43. T. Slama, A. Trevisani, D. Aubry, R. Oboe, F. Kratz, Experimental analysis of an internet-based bilateral teleoperation system with motion and force scaling using a model predictive controller. IEEE Trans. Ind. Electron. **55**(9), 3290–3299 (2008)
44. L. Huijun, S. Aiguo, Virtual-environment modeling and correction for force-reflecting teleoperation with time delay. IEEE Trans. Ind. Electron. **54**(2), 1227–1233 (2007)
45. A. Ramos, D. Prattichizzo, Reaction times to constraint violation in haptics: comparing vibration, visual and audio stimuli, in *Proceeding of the IEEE World Haptics Conference* (2013), pp. 657–651
46. R.H. El Khouli, K.J. Macura, P.B. Barker, L.M. Elkady, M.A. Jacobs, J. Vogel-Claussen, D.A. Bluemke, Mri-guided vacuum-assisted breast biopsy: a phantom and patient evaluation of targeting accuracy. J. Magn. Reson. Imaging **30**(2), 424–429 (2009)

Conclusions and Future Work

This book presented my contribution to the field of haptics and robotics, collecting all the work I have done in the last three years toward my Ph.D. It addresses the challenge of providing effective cutaneous feedback in robotic teleoperation, with the objective of achieving the highest degree of transparency while guaranteeing the stability, and thus the safety, of the considered systems. The book is divided in two main parts: cutaneous-only approaches (Part I) and mixed cutaneous-kinesthetic approaches (Part II).

Part I presented teleoperation systems that provide only cutaneous cues to the operator, thus guaranteeing the highest degree of safety. In fact, cutaneous feedback does not affect the stability of teleoperation systems. We called this approach *sensory subtraction*, in contrast to sensory substitution, as it subtracts the kinesthetic part of the full haptic interaction to leave only the cutaneous cues. In this respect, Chap. 2 presented an application of the sensory subtraction idea in a 1 DoF simulated needle insertion scenario. Sensory subtraction showed intermediate performance between sensory substitution with visual feedback and full haptic feedback provided by a grounded haptic interface. At the same time, sensory subtraction made the teleoperation intrinsically stable, even in the presence of large communication delays. Chapter 3 presented an application of the same cutaneous-only approach in a more challenging remote peg-in-hole task, both in simulated and real environments. Experiments confirmed the results shown in Chap. 2. The sensory subtraction approach showed intermediate performance between no force feedback at all and full haptic feedback provided by a grounded haptic interface. Sensory subtraction again guaranteed the intrinsic stability of the teleoperation system and kept the system stable even in the presence of a communication delay in the haptic loop.

The intrinsic stability of cutaneous-only approaches is extremely promising for those scenarios where safety is a paramount and non-negotiable requirement, such as in robot-assisted surgery. For this reason, in Chap. 4 we extended the results of Chaps. 2 and 3 to a robotic surgery scenario, presenting an application of the sensory subtraction idea in a remote palpation task using the da Vinci Surgical System. Subjects were asked to detect the orientation of a plastic stick hidden in a tissue phantom hearth model. Results showed that providing cutaneous feedback significantly im-

C. Pacchierotti, *Cutaneous Haptic Feedback in Robotic Teleoperation*,
Springer Series on Touch and Haptic Systems, DOI 10.1007/978-3-319-25457-9

proved the task performance in terms of absolute error in detecting the orientation of the plastic stick, completion time, and pressure exerted on the tissue phantom. Moreover, subjects highly preferred conditions providing cutaneous feedback over the one not providing any kind of force feedback.

Although sensory subtraction showed very promising results, providing solely cutaneous stimuli still performed significantly worse than providing the user with full haptic feedback. For this reason, Part II investigated teleoperation systems that provide mixed cutaneous and kinesthetic cues to the operator, with the objective of outperforming the results presented above. In this respect, Chap. 5 presented a teleoperation system with haptic feedback where cutaneous cues are used to compensate for the temporary reduction of haptic feedback necessary to satisfy certain stability conditions. The viability of this approach is demonstrated via one experiment of perceived stiffness and one experiment of teleoperated needle insertion in soft tissue. Results showed improved performance with respect to common control techniques not using cutaneous compensation. Finally, Chap. 6 presented a teleoperation system where mixed kinesthetic and vibrotactile navigation feedback helps the operator in the steering of a bevel-tipped flexible needle in a tissue phantom. Results showed vibrotactile feedback to be more effective than sensory substitution with visual feedback.

The results presented in this book prove cutaneous force feedback to be a viable approach for force feedback in robotic teleoperation. It can substitute haptic feedback to ensure stability, or complement it to improve the system's performance.

In the future I intend to keep working on both research lines. I will study new cutaneous-only approaches, for applications where the safety of the system is paramount, and I will try to understand how to effectively combine cutaneous and kinesthetic cues to improve the performance of existing robotic teleoperation systems. Moreover, I will try to apply these results in new scenarios, such as in micromanipulation, prosthetics, assistive technologies, and rehabilitation.